DIGITAL COMMUNICATION MANAGEMENT

Edited by **Beatriz Peña-Acuña**

Digital Communication Management
http://dx.doi.org/10.5772/intechopen.70959
Edited by Beatriz Peña-Acuña

Contributors

Henrik Vejlgaard, Jana Samáková, Dagmar Babčanová, Henrieta Hrablik Chovanová, Jana Mesárošová, Jana Šujanová, Deepak Verma, Anshu Banwari, Neerja Pande, Maria Da Conceição Da Costa Tavares, Alexander Rozanov, Anna Chebotareva, Vladimir Chebotarev, Teresa Tiago, Tiago Costa, Flavio Tiago, Fawad Kaiser, Ileana Hamburg, Kira Rosa Grosch, Zuzana Birknerová, Miroslav Frankovský, Eva Benková, Ladislav Suhányi, Beatriz Peña-Acuña

Notice

First published in London, United Kingdom, 2018 by IntechOpen
IntechOpen is the global imprint of INTECHOPEN LIMITED, registered in England and Wales, registration number: 11086078, The Shard, 25th floor, 32 London Bridge Street
London, SE19SG – United Kingdom
Printed in Croatia

British Library Cataloguing-in-Publication Data
A catalogue record for this book is available from the British Library

Additional hard copies can be obtained from orders@intechopen.com

Digital Communication Management, Edited by Beatriz Peña-Acuña
p. cm.
Print ISBN 978-1-78923-514-2
Online ISBN 978-1-78923-515-9

For EU product safety concerns:
IN TECH d.o.o., Prolaz Marije Krucifikse Kozulić 3, 51000 Rijeka, Croatia,
info@intechopen.com or visit our website at intechopen.com.

Meet the editor

Beatriz Peña-Acuña is an associate professor at the University of Huelva, Spain.

She enjoys discovering the scientific and human aspects of reality. She is an assistant professor accredited by the Spanish Quality Agency ANECA (2015). Prof. Peña-Acuña graduated in Hispanic Philology and received her diploma in Religious Studies and Journalism. She received the Extraordinary Doctorate Award (2012) with European mention in English. Her scientific production is prolific about communication management and media. Prof. Peña-Acuña directs several collections of books (ACCI, Madrid; MediaXXI, Lisbon; Cambridge Publishing Scholar, UK) and serves as a reviewer of several journals on communication. She has visited several international universities (New York University, UEA, University of London, University of Turku, University of Wroclaw, University of Leiria, University of Sassari, etc.). Prof. Peña-Acuña organizes international conferences. She received the President'sLifetime Achievement Award (2015) and Honoris Causa of UFA University, Russia (2016).

Contents

Foreword

The need for the management of communication in the company is shown by the presence of a communication department and a communication manager (DirCom). This depends on the executive command or the executive control. This need has been understood for a decade in the communication trend in Europe.

The European Communication Monitor 2017 (by Ansgar Zerfass *et al.*) is titled "How strategic communication deals with the challenges of visualisation, social bots and hypermodernity".

This book highlights what we have pointed out above on the strategic business question: the long-term development of strategic issues for communication management in Europe, consisting of two main issues at the top of the list: linking business strategy and communication, and coping with the digital evolution and the social web.

This book tells us about external communication across Europe, where social media and social networks are considered by far the most important channel to address stakeholders, gatekeepers and audiences. Indeed, coping with the digital evolution and the social web has been voted as the single most important issue for communication management over the next three years. It also points to this interesting fact about internal communication: the communication function emerging as a key supporter for daily management and the operations of other departments.

It also says that an overwhelming 94.4% of European communication professionals believe that visual communication will gain in importance for organizations and only one out of 10 communicators rate themselves as highly skilled in visual communication. Strategic communication has always involved a range of media platforms and visualisation has been frequently at the centre of message transfer. As technology continues to evolve, visual support for communication messaging has increased and diversified. The rapid growth of social networks such as Facebook, YouTube and Instagram has inevitably created an increase in the use and application of visual stimuli in these formats.

The second edition of the Asia-Pacific Communication Monitor (2017–2018) focuses on social media, strategic challenges and professional capabilities. The Latin American Communication Monitor (2016–2017) studies the professional competencies and changes.

The panorama of communication management is transformed very quickly so that until recently, we talked about internal communication 2.0. But now we talk about co-created internal communication and innovation through computer media such as Design Thinking. It is necessary to put into play as an added value leadership and human resources in its most human facet and its communicative potential capable of improving productive and creative processes, and also achieving a climate of self-confidence and a culture that reinforces teamwork.

Alejandro Formanchuk,
Formanchuk and Associates,
Buenos Aires, Argentina

Preface

Companies understand that an intelligent communication direction is translated into value for the company where the asset, that is, people, can be involved and flow in production processes with higher quality, the culture and values of the company are strengthened. This directly affects the productivity of the company and its economic performance. The most successful companies are those that have a human fabric and a consolidated business culture.

Data from the study on DirCom in Spain 2015 offers us the following insights. The role of communication is growing. More than 80% of the respondents estimate that the communication of companies has increased its importance and its role, which are located at the first executive level. The communication department is consolidated as an area of relevance in companies, depending on the first executive level. The dependence of the marketing management is progressively reduced. The management of companies is more involved in defining the communication strategy of the companies.

The director of communication is a profession that is very zealous to discover which strategies are the best and the most intelligent. There are not many manuals and there are some that offer general and little updated information about the change that new technologies imply. We find that this literature is isolated from big data and from social media, which can be directly useful. The consultants prefer to give training to employers, and this training is extended to middle managers and employees on site. These trainers usually publish little, except for some, who describe their formulas, approaches or their models. However, we will say that there is no single recipe for DirCom or communication consultants. Each one will offer models according to the variables or factors that he/she thinks will rectify the direction of a company or the preparation of communication by a CEO according to his/her personal mood. After dedicating many years of study to business cases, we think that communication strategies must be sought for each company, planned and executed for each individual. In this way, the adaptation can be more adjusted. Even so, we emphasise the importance of internal communication strategies, a strategy that may not involve economic investment.

Thanks to the large studies on communication management, we can understand such complex processes that are changing fast because of new technologies and new trends . In this book, we mention the research by Zerfass and other academics in European Communication Monitor (communicationmonitor.eu), in Asia Pacific Communication Monitor (zerfass.de) and coming soon in North American Communication Monitor (plankcenter.ua.edu), and the Latin American Communication Monitor led by Maria Angeles Moreno and Juan Carlos Molleda (latincommunicationmonitor.com). There are other interesting studies about Internet usage worldwide (statista.com) and studies about communication management in some countries (as the Study of Communication in Spain 2015 (dircom.com)).

The trends focused on by the European Communication Monitor in 2017 are visuality, social (ro)bots, and hypermodernity. One trending area in communication management is the use and application of software robots, or what we most frequently refer to as 'social bots'. Social bots have been heavily used and have probably influenced political election outcomes such as during the United States presidential campaign in 2016 as well as during the BREXIT vote in the United Kingdom.

Theoretically, the current global society can be labelled as a hypermodern society. Hypermodernity is a concept introduced by the French social theorist Gilles Lipovetsky (2005).

A hypermodern society is a society in overdrive, characterised by a culture of hyperconsumption, hyperchange and hyperindividualism. This study, for the first time, asked European practitioners about aspects of hypermodernity and how this influences their organization(s) and their work. About 43.5 per cent of the organisations were already changing from postmodern to hypermodern with characteristics such as continuous change, decentralised IT, rapid adjustments of the workforce, creativity and ethics of perceived responsibility.

We understand that visuality is in fact a predominant language in the culture in general and that it has also been established in the communication of companies as a tool in internal as well as external communication. The study indicates that almost all communicators believe in an increasing importance of visualisation for strategic communication, but competences and processes are still incipient.

Among the visual elements that have been taken into account in this book, are the following ten elements: online videos (for example, web clips), art (for example, paintings and abstract photos), space design (for events and rooms), professional films (for example, picture films and commercials), professional photos (prearranged/edited, in stock), online animations (for example, flash and web banners), signs and symbols (for example, logos, icons and pictograms), commercial graphics (for example, tables and figures), snapshots (spontaneous, unedited), and infographics (for example, explanatory content).

The direction of communication is a profession that increasingly requires an update because it is a scenario with a rapid change due to the demands of new technologies and the predominance of these compared to traditional media. This changing picture is shown by the latest studies of the European Communication Monitor since 2013. So, today, we can speak of Digital Communication Management.

Beatriz Peña Acuña
University of Huelva
Huelva, Spain

Section 1

Digital Communication Management

Chapter 1

Introductory Chapter: Changing Panorama of Digital Communication Management

Beatriz Peña Acuña and Alejandro Formanchuk

Additional information is available at the end of the chapter

http://dx.doi.org/10.5772/intechopen.78071

1. Introduction

The need for the management of communication in the company, that there is a Department, a communication manager (DirCom) prepared in Communication, that depends on the executive command or that is even in the executive control, is understood for a decade in the Communication trend in Europe.

Companies have understood that an intelligent communication direction is translated into a value for the company where the asset, that is, people can be involved and flow in production processes with higher quality, the culture and values of the company are strengthened. The company and, of course, directly affects the productivity and economic performance of it. The most successful companies are those that have a human fabric and a consolidated business culture.

The data of the study on DirCom in Spain 2015, for example, offer us the following data. The role of Communication grows. More than 80% of the respondents estimate that the Communication of the companies has increased its importance and its role, which are located in first executive level. The Communication Department is consolidated as an area of relevance in companies, depending on the first executive level. The dependence of the Marketing Management is progressively reduced. The management of companies is more involved in defining the communication strategy of the companies.

The director of communication is a profession that is very zealous to discover which strategies are the best and the most intelligent. There are few manuals and there are some that offer general and little updated information about the change that new technologies imply. We find literature isolated from big data, from social media that can be directly useful. The consultants prefer to give training to employers, and extended to middle managers and employees on site. These usually publish little, except for some, who diffuses what are their formulas or approaches or their models. However, we will say that there is no single recipe for DirCom or Communication consultants. Each one will offer models according to the variables or factors that seem to them to be able to rectify the direction of a company or the preparation in communication of a CEO according to his personal mood. After dedicating many years of study to business cases, we think that communication strategies must be sought for each company, planned and executed for each one. In this way, the adaptation can be more adjusted. Even so, we emphasise the importance of internal communication strategies, a strategy that may not involve economic investment.

2. Trends in the digital communication management

The European Communication Monitor 2017 (by Ansgar Zerfass et al.) is titled *How Strategic Communication Deals with the Challenges of Visualisation, Social Bots and Hypermodernity.*

This study highlights what we have pointed out above on the strategic business question: The long-term development of strategic issues for communication management in Europe, consisting of two main issues at the top of the list: linking business strategy and communication; coping with the digital evolution and the social web.

This study tells us about external communication that across Europe, social media and social networks are considered by far the most important channel to address stakeholders, gatekeepers and audiences. Indeed, coping with the digital evolution and the social web has been voted as the single most important issue for communication management over the next 3 years. It also points this interesting fact about internal communication: the communications function emerging as a key supporter for daily management and the operations of other departments.

It also says that an overwhelming 94.4% of European communication professionals believe that visual communications will gain in importance for organisations, only 1 out of 10 communicators rate themselves as highly skilled in visual communications. Strategic communication has always involved a range of media platforms and visualisation has been frequently at the centre of message transfer. As technology continues to evolve visual support for communication, messaging has increased and diversified. The rapid growth of social networks such as Facebook, YouTube and Instagram has inevitably created an increase in the use and application of visual stimuli in these formats.

One trending area in communication management is the use and application of software robots, or what we most frequently refer to as 'social bots'. Social bots have been heavily used and probably influenced political election outcomes such as during the United States presidential campaign in 2016 as well as during the BREXIT vote in the United Kingdom.

Theoretically, the current global society can be labelled as a hypermodern society. Hypermodernity is a concept introduced by the French social theorist. A hypermodern society is a society in overdrive, characterised by a culture of hyper consumption, hyper change and hyper individualism. This study, for the first time, asked European practitioners about aspects of hypermodernity and how this influences their organisation(s) and their work. About 43.5% of the organisations are already changing from postmodern to hypermodern with characteristics such as continuous change, decentralised IT, rapid adjustments of the workforce, creativity and ethics of perceived responsibility.

The Direction of communication is a profession that increasingly requires an update because it is a scenario with a rapid change due to the demands of new technologies and the predominance of these compared to traditional media. This changing picture is shown by the latest studies of the European Communication Monitor since 2013. So, we can speak today of the Digital Communication Management.

Author details

Beatriz Peña Acuña[1]* and Alejandro Formanchuk[2]

*Address all correspondence to: bpena@uhu.es

1 University of Huelva, Spain

2 Formanchuk and Associates, Buenos Aires, Argentina

Chapter 2

Process Knowledge in the Innovation-Decision Period

Henrik Vejlgaard

Additional information is available at the end of the chapter

http://dx.doi.org/10.5772/intechopen.73307

Abstract

Current diffusion of innovations theory states that knowledge is a key first stage in the adoptive process of an innovation. However, the fact that different types of knowledge may be relevant to different adopter categories does not appear to have been investigated. In this study, the concept of process knowledge is introduced into the adoptive process of an innovation. The study object was digital terrestrial television (DTT). Data were gathered at eight data gathering points in an 18-month study period. Three different types of process knowledge were identified at different stages: the help knowledge stage, the customer participation knowledge stage, and the interaction knowledge stage. In this study, it is suggested that the following three questions are the ones that majority adopters and laggards want to be answered in the knowledge stage: (1) What is the innovation? (2) What do I need to do to adopt? and (3) Who can help me in the adoptive process? With answers to these questions, consumers have the knowledge that may help speed up the rate of adoption of an innovation. This has practical implications in communication management, for instance, for change agents who are framing messages.

Keywords: diffusion of innovations, rate of adoption, digital terrestrial television, majority adopters, laggards

1. Introduction

Knowledge plays an important part in the diffusion of an innovation as it represents a key stage in the innovation-decision period, the period from when a consumer has awareness-knowledge of an innovation until he or she has adopted the innovation. While innovations, at least at a certain time in history, may have been perceived as mainly technological, innovations in fact can have many different manifestations. One can add to that that an innovation, whether it is technological or not, may have social ramifications that makes the diffusion of the innovation a key societal priority. This can be the case with respect to digital technology, which is likely to be utilized more and more in both government and nongovernment communications. This makes knowledge on how to speed up the diffusion process, especially with respect to digital laggards, an important issue. In this communication management process, framing of messages plays a key role.

Knowledge on the diffusion of an innovation is just one aspect of diffusion of innovations research, much of which was carried out in the twentieth century. The outcome of this research has resulted in a scientific theory of diffusion of innovations, that is, a theory that has been validated by scientific research, with Everett M. Rogers' book *Diffusion of Innovations* [1] representing the scientific theory of diffusion of innovations (up till manuscript completion of the latest edition of his book, published in 2003). Scientific theory is opposed to analytical theory, which has not been validated by science, but may be based on abductive research, in the understanding of science philosopher Charles Sander Peirce [2]. While all aspects of the diffusion of innovations certainly cannot be said to have been fully investigated, *Diffusion of Innovations* is an example of research becoming scientific theory. It is worth noting that according to a literature review of diffusion of innovations academic literature in the period 2002–2011, "diffusion research seems to be data driven and relies heavily on empirical data that support and change theories in modest ways only" [3]. This appears to be the case also with the diffusion of innovations research that was carried out before this specific review.

What is worth noting is that this specific theory is based on research that was carried out in the twentieth century, with some of the research of key aspects of the theory having been carried out in the period from the 1940s to the 1960s. By elevating research from the mid-twentieth century to scientific theory, we also implicitly accept that the research that was carried many years ago still has validity. This is a philosophy of science issue not only within the field of philosophy of science but also within the specific discipline or field of study. This paper will address this issue from the perspective of one field of study, namely, diffusion of innovations.

According to Charles Sander Peirce, science may result from different scientific processes. Peirce himself reexamined the three basic modes of inference—abduction, deduction, and induction—originally introduced by Aristotle [4]. Each of the three modes of inference has a specific and logical purpose according to Peirce: in abduction, a theory is formulated, based on casual observations. Abduction can be viewed as qualified guesswork (a term also used by Peirce [5] or as creating a theory based on what seems to be cognitively logical. A theory that is the outcome of abduction may become part of scientific discourse, without being empirically tested. But from the theoretical understanding generated by the abductive approach, other studies can be undertaken to validate or falsify the theory by way of deduction or induction. In Peirce's three-way approach, the abductive theory should be tested deductively. If falsified, the theory must be discarded (cf. [6]). If verified, it becomes science (scientific theory). In Peirce's scientific approach, a confirmed theory must be tested continuously through deduction and/or induction; in this way, a theory can stay updated or real. Likewise, even though well established, a scientific theory must be challenged on a continuous basis. One can argue, quite banally, that when society changes then one can expect theories about society also to evolve, to change, to become outdated, or to lack nuances. This is also the philosophical approach to the subject matter of this paper.

While a theory may be well established as scientific theory, it may be challenged for other reasons as well. With respect to diffusion of innovations, which is a hybrid of the communication and sociology disciplines, there are issues with general or universal validity, especially because much early research was carried out in a certain cultural context ("the Western world"). One can add to that that a major part of the research was carried out in a time with less technology, certainly without digital technology, compared to what we see today. One can, therefore, argue that there are ample reasons to continue to seek validation of the diffusion of innovations theory.

In the tradition of Peirce, this paper will investigate the knowledge stage in the innovation-decision period, utilizing data that was gathered in the beginning of the twenty-first century when digital communication was fairly well established. However, to further underline the digital age context, the study object is a digital service, namely digital terrestrial television (DTT).

The philosophical approach of this paper intentionally focuses, and limits, the scope of the paper, as the aim of the paper is specifically to nuance one aspect of diffusion of innovations theory in a twenty-first-century applied communication context. Therefore, a brief summary of key aspects of diffusion of innovations theory is included below, based on Rogers' theorizing [1].

2. Diffusion of innovation theory

Diffusion is "the process by which (1) an *innovation* (2) is *communicated* through certain *channels* (3) over *time* (4) among the members of a *social system*" ([1], p. 11). Diffusion is also termed the adoptive process in which the rate of adoption is a key concept. The rate of adoption is defined as "the relative speed with which an innovation is adopted by members of a social system" ([1], p. 23). It is generally measured as the number of individuals who adopt an innovation in a specified period, such as a year ([1], p. 221). When the number of individuals adopting a new idea is plotted on a cumulative frequency basis over time, the resulting distribution is an S-shaped curve ([1], p. 23). The slope of the S-curve typically rises slowly at first and then steadily before it flattens. However, the slope of the S-curve can take many forms ([1], p. 328). The steeper the slope of the S-curve is, the faster the rate of adoption is.

Some innovations diffuse rapidly, and the S-curve is then quite steep. Some innovations have a slower rate of adoption which makes the curve more gradual. The rate of adoption is typically measured by the length of time required for a certain percentage of the members of a system to adopt an innovation. "Therefore, we see that the rate of adoption is measured for innovation in a system, rather than for an individual as the unit of analysis. […] This system may be a community, an organization, or some other structure" ([1], p. 23). The members or units of the social system may be individuals, groups, or organizations. Thus, the social system can, for instance, be a segment of consumers sharing the same trait(s) or, indeed, all consumers in a country ([1], p. 24).

The S-curve illustrates that people have different approaches to innovations: some people adopt innovations right away, others need longer time, and some people need a very long time.

This led Rogers to categorize people by adopter categories, the classifications of members of a social system on the basis of innovativeness ([1], p. 22). He identified five adopter categories: innovators, early adopters, early majority, late majority, and laggards.

The adopter categories have different personalities, interests, financial circumstances, and educational levels. Individuals who are innovators are different from the late adopters. Innovators are more open and curious than late adopters, and they want an innovation as soon as they hear about it. Laggards are the opposite and take the longest time to adopt. With respect to laggards, Rogers wrote that they "are near isolates in the social networks of their system [...] [and] tend to be suspicious of innovations and change agents" ([1], p. 284).

The innovativeness dimension is used to understand and define the five adopter categories in quantitative terms. This dimension is measured by the time at which an individual adopter category adopts an innovation. Rogers partitioned the adopter categories into the five categories by calculating the standard deviation from the average time of adoption. This gave the following result ([1], p. 281): (a) innovators, 2.5%; (b) early adopters, 13.5%; (c) early majority, 34%; (d) late majority, 34%; and (e) laggards, 16%.

In diffusion of innovations theory, knowledge is an individual's initial exposure to the innovation's existence and understanding of how the innovation works ([1], pp. 177–174). There are three types of knowledge that all belong to the knowledge stage in the innovation-decision process, with the knowledge stage being the first of the five stages in the innovation-decision process: knowledge, persuasion, decision, implementation, and confirmation ([1], pp. 171–173). "The innovation-decision process is an information-seeking and information-process activity in which an individual obtains information in order to gradually decrease uncertainty about the innovation. [...] At this stage the individual wants to know what the innovation is and how and why it works" ([1], pp. 20–21). According to the diffusion of innovations theoretical framework, the answer is the three types of knowledge in the innovation-decision period: (1) awareness-knowledge, (2) how-to knowledge, and (3) principles-knowledge. All the three types of knowledge belong to the knowledge stage in the innovation-decision process ([1], pp. 171–173). According to Rogers, each of the three types of knowledge can be framed as a question: (1) What is the innovation? (2) How does it work? and (3) Why does it work? The first question is about the existence of an innovation; the second question is about knowledge necessary to use an innovation properly; and the third question is knowledge about the functioning principles underlying how an innovation works ([1], p. 21). These questions can be characterized as innovation-centric. It is worth pointing out that much diffusion of innovations research does appear to have a pro-innovation bias ([1], pp. 106–107), and this may also be the case with respect to the knowledge stage of the innovation-decision process.

It is the innovation-decision process that leads to either adoption or rejection, a decision not to adopt an innovation ([1], p. 21). Adoption takes place at the decision stage of the innovation-decision process ([1], p. 170). Only when having knowledge about the innovation can an individual be persuaded to adopt. It may be easier to persuade an individual to adopt if the individual has knowledge that he considers relevant. What the individual considers relevant may be depending on the "Characteristics of [the] Decision-Making Unit" (ibid.), that is, the individual's adopter category profile.

One could speculate that the above types of knowledge, in particular type nos. 2 and 3, will mainly have appeal to innovators. The innovators have different characteristics than the other adopter categories, and they are the keenest to adopt ([1], pp. 287–292), hence an interest in innovation-centric knowledge. However, other types of knowledge may appeal to other adopter categories in *their* innovation-decision process. This study will investigate if other types of knowledge can play a role in the innovation-decision period, especially knowledge that can influence the innovation-decision behavior of late adopters, in particular laggards who take the longest time to adopt ([1], p. 215).

3. Literature overview

One article has identified more than 5,000 articles using diffusion of innovations theory [7]. Many articles use other models to explain adoption, for instance, the technology acceptance model [8], which focuses on technology and represents a different perspective than the diffusion of innovations theory utilized in this study. It should also be noted that knowledge as part of the adoption process is not exclusive to diffusion of innovations theory (see, for instance, [9]). Across theoretical approaches, it seems that some degree of knowledge regarding the existence, uses, and meaning of the innovation can influence the adoption decision. This knowledge can be actively sought by end-users, and/or it can be communicated by one or more change agents ([1], Chapter 9). However, it is notable that relatively little interest appears to have been devoted to the knowledge-type aspect of the innovation-decision stage in recent years, even though knowledge and access to knowledge seem to be undergoing many changes, not least because of rising educational levels, technology, and (social) media proliferation.

Within the context of this study, it is also worth noticing that the knowledge aspect has been specifically addressed in three studies involving terrestrial digital television [10–12]. In an early study of audience interest in adopting digital television (DTV), it was pointed out that "we still know relatively little about viewer knowledge about and interest in adopting the new, higher resolution television receivers [necessary to receive DTT]" [10]. It was concluded that "the fact that that fewer than a third of respondents feel even "somewhat educated" about DTV—only a few years before its mandated adoption in the United States—is remarkable" [10]. The findings of the study opened up questions as to what type of knowledge could help facilitate the adoption of DTT.

Dupagne [11] measured self-reported DTV knowledge, also in the United States, and pointed out that this is "not actual DTV knowledge. [...] An individual may well report a degree of familiarity with DTV but may nevertheless misconstrue how the technology functions or how to use it." The author is here de facto referencing, respectively, knowledge type nos. 3 and 2. It does not appear obvious that how-to knowledge and principles-knowledge will help facilitate the adoptive process if technology insight is minimal and there is no real interest in the technology aspect of the innovation.

A third study found that knowledge of DTV had a significant influence on the intention to adopt DTV. This study also pointed out that a lack of knowledge was likely to lead to a delay in making any decision on adopting DTV [12]. The study categorized DTV knowledge into three categories: environment, content, and equipment. These categories appeared after a review of the frequently asked questions reported by American DTV information websites. The environment knowledge category (which is mainly knowledge about process) can include statements like, "DTV sets are in stores now" or "There is a government deadline for DTV conversion." The content knowledge category can include statements like "DTV enables interactivity on TV" and "DTV enables multicasting." The equipment knowledge category can include statements like "Regular TV sets will be obsolete when only digital transmission is in the air" and "A converter is needed to receive DTV signals." The findings showed that, among the three knowledge categories, the DTV environment and DTV content categories seemed to be the most relevant to the adoption intention of new TV sets and that the DTV environment knowledge category had significance with respect to the adoption intention of converters. It seems that DTV environment knowledge, in particular, may have an effect on innovation-decision process. The three knowledge-type categories that currently are proposed as key in the innovation-decision process are innovation-specific, and, especially, 2. How-to knowledge and 3. Principles-knowledge appear to belong in the DTV equipment knowledge category. This type of knowledge may not have the effect on the innovation-decision process that theory suggests.

It should be noted that studies of knowledge are not only of the stages or of the process. Also, content is studied (see, for instance, [13]), as are the psychological processes underlying knowledge (knowledge structures) [14]. Several studies have looked at the barriers to adoption of technology, of which lack of knowledge may play a part. A meta-analysis of factors determining older adults' technology adoption addresses some of the same issues as the present study [15]. Lee and Coughlin write "The factors suggest that older adult's adoption of technology is not a purely technical topic, but a rather complex issue with multiple aspects" ([15], p. 750). The factors include delivery channels, defined as "Ways in which technology is communicated and distributed to older adults for purchase and use" ([15], p. 750). An earlier study had suggested that older adults' access to technologies relies on how much information is open to them and how the delivery systems are formed [16]. This could indicate that knowledge about the distribution channel, for instance, stores, can be relevant.

In conclusion, it appears that the previous research indicates that it can be relevant to explore environment knowledge in the innovation-decision period. Knowledge related to questions, like How does it work? and Why does it work?, may be interesting to early adopters. The late adopter categories may be more interested in environment knowledge questions like "Where can I get information?", "Who can help me?", and "Which actions do I have to take in order to adopt?" These types of questions have a different perspective, in that they reflect that the adopter categories being asking these questions may not be interested in the innovation as such, only in the outcome of the adoption process, for instance, being able to watch television or use digital technology.

The three questions mentioned above reflect that maybe an environment knowledge perspective could help facilitate the innovation-decision process: especially, majority adopters and laggards may be more prone to adopt if they know where they can get help and know how to ask for help. However, environment knowledge may not be the right knowledge category term, as the word environment may lead to associations to environmental issues. Therefore, it is suggested to use another term that precisely describes what this knowledge category is about, namely, knowledge about process. Instead of using the term environment knowledge, the term process knowledge will be used. In a study of lead users, a distinction is made between product knowledge, process knowledge, and knowledge about factual matters [17], confirming the meaningfulness of the term in this context. However, one should be aware that in other contexts, the term process knowledge may have slightly different meanings. For instance, in the study of text mining, process knowledge means specific documents that are used in many processes, for instance, communication logs and process descriptions [18]. In the present study, process knowledge does not relate to documents.

4. Research question

The research question is as follows:

> RQ: In a diffusion of innovation process, how high are process knowledge levels when majority adopters and laggards adopt an innovation?

Three types of process knowledge are selected for this study: (A) knowledge on where to get help, (B) knowledge about the actions required by the user/customer/consumer in order to adopt, and (C) knowledge on how to interact with staff at a store.

In order to answer the RQ, data on the rate of adoption of majority adopters and laggards has to be established. This de facto means that the entire rate of adoption of the innovation has to be established, as laggards are the last to adopt.

The units of analysis of the study are Danish households with analog terrestrial television (ATT). This segment is the approximately 600,000 Danish households with ATT as of March 31, 2006 ([19], p. 319).

In June 2005 the Danish Parliament decided that the ATT signal was to be switched off permanently on the night between October 31 and November 1, 2009. The digital terrestrial television signal was "in the air" in Denmark for the first time on March 31, 2006, and would be so along with the ATT signal until the cutoff date. After the cutoff date, only the DTT signal would be in the air [20, 21]. In other words, the entire adoptive period is from March 31, 2006, to October 31, 2009.

Knowledge is a variable in the innovation-decision process. Knowledge, ultimately, has an effect on the rate of adoption. However, a person does not get knowledge without proceeding stimuli, typically interpersonal communication and/or mass communication, both of which can take place with a change agent as the sender. It is well known from diffusion of innovations theory and research that many different variables determine the rate of adoption ([1], Chapters 5 and 6). Therefore, in the adoptive process, the variable representing knowledge is in fact an intermediate variable, as there were many public information and commercial are activities informing of the ATT switch-off affecting the knowledge of the consumers. In this study, the campaign activities are rendered a black box, but they have been described by Sepstrup [19]. There are also typical prior conditions, for instance, felt needs/problems ([1], p. 170). These variables are interconnected in the following way:

Prior conditions (conditioning variable) → communication stimuli (causal variable) → knowledge (intermediate variable) → rate of adoption (effect variable)

In diffusion of innovations theory, knowledge is presumed to be a de facto active causal variable in a very complex process.

5. Methodology

In classic diffusion of innovations research, it was common to use data based on respondents' memories, sometimes going back 10–15 years. Much of the original diffusion research gathered data from adopters by asking respondents to look backward in time after the innovation had diffused ([1], pp. 126–127). Rogers suggested an alternative research approach to the after-the-fact data gathering. He wrote "It is possible to investigate the diffusion of an innovation while the diffusion process is still underway. [...] Data can be gathered at two or more points during the diffusion process, rather than just after the diffusion process is completed [...]" ([1], p. 112). The question of diffusion of innovations methodology, and the problems with the early research methods, has been addressed by Meyer [22] who also suggested, among other innovative methodologies, point-of-adoption studies. In point-of-adoption studies, "data is gathered from respondents at the time they adopt the innovation rather than at some point in the distant future. [... enabling] the investigator to obtain more accurate data about the innovation decision [...]" [22]. This paper represents a point-of-adoption study. In order to measure the affective variable, the research design is based on conducting a series of surveys over a period of time, at the time when the adoptive process is taking place (thus, this study is not based on respondents looking backward in time).

Rogers indicated that three points at which data are gathered can form an S-curve if you also have a zero ([1], p. 113). In this study there were eight data gathering points. Eight data gathering points should thus yield a very "readable" curve.

A structured interview guide consisting of 72 questions was used for all the respondents. The questions were of a sociodemographic nature and about TV habits and TV reception. For this study the following three survey questions were asked, each representing a different type of knowledge:

Survey question A: Do you know where you can get further information?

Survey question B: Do you know how you ensure that you can watch television after the analog signal has been switched off?

Survey question C: Did you feel well prepared for the encounter with the store?

To establish the rate of adoption, the following survey question was asked:

Survey question D: Has your household converted to the digital terrestrial television signal?

In a structured interview with 72 questions, one should very carefully consider the reply categories and aim to keep them consistent, yet meaningful to each question. While scale measuring is often considered the most suitable way to elicit replies from respondents by the researcher, the researcher should also try to view the interview from the point of view of the interviewee: when carrying out structured interviews with fixed reply categories, one should carefully consider the use of scales and in all circumstances limit the use of scales to as few as possible. As the interviewee only hears the scale spoken, he may be more unsure of the scale than when viewed on paper or on a screen. Some scales may result in a high number of replies; others will result in very precise replies. The researcher wants to achieve both. In this study, it was decided that the reply categories for all four questions should be the same and that three reply categories could satisfactorily represent the expected replied to the questions. The reply categories for all four questions were "Yes," "No," and "Don't Know," which appear to be reply categories that are likely to be meaningful to the respondent because of the way the questions are phrased. With respect to survey question D, it was possible to cross-check and to verify responses concerning television reception by asking multiple behavioral and technical questions. Thus, if someone answered "I don't know" to the survey question *after* the ATT switch-off, it would in fact be possible to change this to a "Yes" or a "No" in a simple way: ask the respondent to turn on the television set, and see if there is a signal. If the answer to that question was a "Yes," the answer to survey question D would also be a "Yes."

Adoption could not start to take place until the DTT signal was in the air on March 31, 2006. Therefore, March 2006 is set as zero.

6. Data gathering

As a consequence of the above, eight surveys were carried out, at the following times: June 2008, October 2008, January 2009, June 2009, August 2009, September 2009, October 2009, and November 2009. The precise data gathering process has been reported in previous research, utilizing other data from the data sets than what are reported here. Therefore, the following is a description of a data gathering process that has been reported in other academic papers (for instance, [23, 24]).

The June 2008 survey was carried out in a different way from the subsequent seven surveys. The June 2008 survey was an internet-based questionnaire with close-ended questions distributed to a representative number of Danes aged 18+, with their own household, from a large base of potential respondents, selected randomly. The number of respondents was 969 individuals. The survey question was the same as for the subsequent surveys. The respondents were representative of the Danish population on all standard parameters. The response rate and statistical variance could not be established for this survey. This is an obvious problem with respect to reliability, and this should be borne in mind when analyzing the data, especially if the data are not "in line" with the subsequent survey results.

The seven surveys carried out in October 2008–November 2009 were all done using the same method. Surveying television reception can be carried out in many ways. It is important to be aware of the technological issues involved and of the fact that some consumers have little or no knowledge of their own television reception.

For each of the seven surveys, the following method was used: a sample population representative of the Danish population, aged 18+, with their own household, was selected randomly. After selection the potential respondents received a letter informing them that they had been selected to participate in a survey of television habits. Each person in the sample population was then contacted in one of two ways: 80% were contacted by telephone and 20% were visited in their own home by an interviewer. Irrespective of contact method, the respondent selected for interviewing was the household member responsible for the TV equipment in the household. If a potential respondent could not be reached or refused to participate, a new potential respondent with the same sociodemographic characteristics was selected and included in the sample. Thus, it was secured that the sample population would always be representative. Consequently, the surveys in principle can be said to have 100% response rates. The selection and interviewing process lasted 4 weeks.

7. Survey results

The results of the eight surveys are shown in **Table 1**.

	March 2006+	June 2008	October 2008	January 2009	June 2009	August 2009	September 2009	October 2009	November 2009
Number of Respondents	-	969	1001	1024	978	977	989	998	976
A. Do you know where you can get further information?		*	43%	44%	55%	59%	60%	**	***
B. Do you know how you ensure that you can watch television after the analogue signal has been switched-off?		*	87%	92%	81%	84%	78%	**	***
C. Did you feel well prepared for the encounter with the store?		*	*	65%	74%	87%	90%	**	80%
D. Has your household converted to the digital terrestrial television signal?	0%	4%	*	13%	**	**	26%	38%	99,7%

As a general rule numbers are rounded up or down.
+ DTT launched in Denmark March 31, 2006 (not a survey).
*Not part of this survey.
** Not reported because of statistical variance due to the error level compared to the January 2009 survey.
*** Not included in the survey because of too few respondents (under 31).

Table 1. Survey questions results. Danish households affected by the ATT switch-off.

Survey Question C	Result	Result with Statistical Variance
January 2009	13%	0-32%
September 2009	26%	0-43%
October 2009	38%	22-54%

Table 2. Statistical variance for survey question D.

For questions A, B, and C, the statistical uncertainty for all surveys was ±3.6% or less. For question D, the statistical uncertainty is reported in **Table 2**.

8. Findings

With the survey results, the RQ can be answered. Three types of process knowledge were measured among the affected households: (A) knowledge on where to get help ("help knowledge"), (B) knowledge about the actions required by the user/customer/consumer in order to adopt ("customer participation knowledge"), and (C) knowledge on how to interact with staff at a store ("interaction knowledge").

Help knowledge had the lowest knowledge level in the entire process. From 43%, the number grew 17% points. Customer participation knowledge started out with a high knowledge level in the period measured: From 87%, the knowledge level fell 9% points (at the second measuring point, it rose by 5% points, and then it started falling). Interaction knowledge rose during the measured period up to the deadline but fell at the very end of the adoptive period. From 65%, it rose 25% points (at the very end, it fell 10% points).

It appears that at different stages in the adoptive process the level of certain types of knowledge is high; in fact, there appear to be three different process knowledge stages: the help knowledge stage, the customer participation knowledge stage, and the interaction knowledge stage. The help knowledge stage was 11 months, from October 2008 to September 2009. Of the three knowledge stages, the help knowledge stage had the lowest average knowledge level (52%), that is, of the three types of process knowledge, most adopted without having this type of knowledge.

The customer participation knowledge stage was the 7 months' period from June 2008 to January 2009 in which period customer participation knowledge was at the highest. This stage had the highest average knowledge level (84%). The interaction knowledge stage was from September 2009 to November 2009 in which period interaction knowledge was at the highest. This stage had the next-highest average knowledge (79%). The customer participation knowledge stage and the interaction knowledge stage have the highest average knowledge levels of the three process knowledge categories.

In order to answer the RQ, the rate of adoption has to be established. The rate of adoption of DDT in Denmark was 99.7% which is here considered as 100% rate of adoption; in other words, both the majority adopters and the laggards have adopted DTT. However, just 1 month before

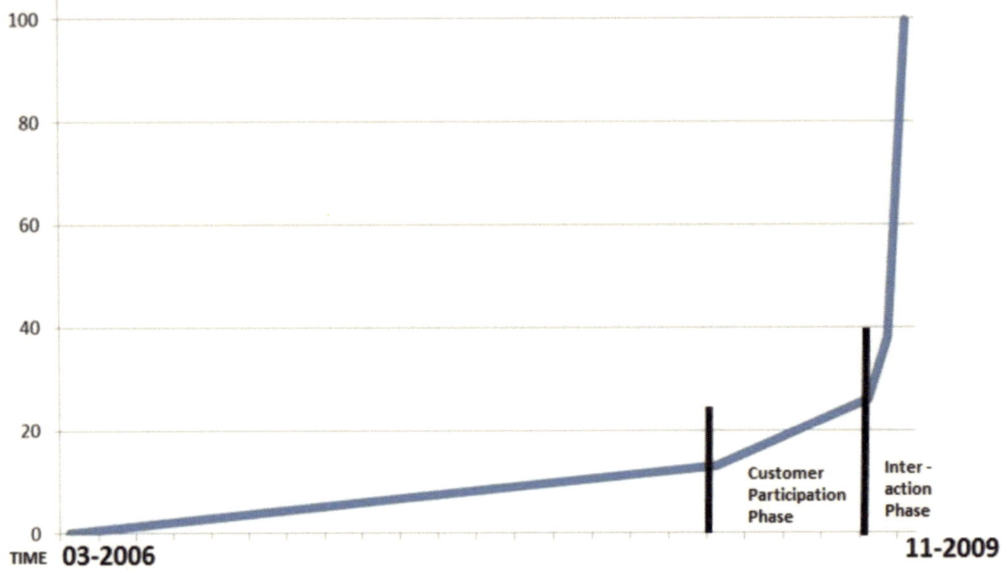

Figure 1. The rate of adoption of DTT in Denmark with two process knowledge stages.

the deadline, only 50% had adopted. The curve at the end of the adoptive process is extremely steep which means that the rate of adoption at this stage was very fast, making it an atypical S-curve. However, even with this atypical S-curve, it is possible to establish the five adopter categories' time of adoption. The 15–84% range is when the majority adopters adopt, and the 16–100% range is when the laggards adopt. After the 15% mark, there is an upward movement in the curve, indicating that when the early majority adopters adopt customer participation knowledge is high. After the 25% mark, there is an upward movement in the curve, indicating that when many majority adopters and laggards adopt, interaction knowledge is high. The upward movements in the curve show that the rate of adoption becomes faster.

In **Figure 1**, the numbers from survey question D in **Table 1** have been turned into a curve. In the figure the customer participation knowledge stage and the interaction knowledge stage are identified. These two stages represent the highest process knowledge levels when the majority adopters and the laggards adopted.

9. Discussion

This paper has investigated what type of knowledge is relevant to consumers in the innovation-decision period, with a focus on the application in communication management. In established diffusion of innovations theory, the adopter categories need answers to the following three questions: (1) What is the innovation? (2) How does it work? and (3) Why does it work? While these three questions and their answers may play a key role for innovators and early adopters in *their* innovation-decision process, here it is suggested that other questions and issues may be more relevant and more meaningful to majority adopters and laggards.

It seems that process knowledge can play a part in the diffusion of innovations process in the middle and the late part of the adoptive process. While question type no. 1, What is the innovation?, may be relevant to all adopter categories, it could appear that the two other questions that late adopters may want answers to are *not* equipment related. Rather, they are process related: knowledge on which actions are required in order to adopt and about how to interact with someone who can help (for instance, store staff). Based on the findings in this study, it is suggested that the following three questions are the ones that those who adopt *after* the innovators and early adopters have adopted want answered: (1) What is the innovation? (2) What do I need to do to adopt? and (3) Who can help me in the adoptive process? With answers to these questions, consumers have the knowledge that will help speed up the rate of adoption. This insight may be relevant with respect to many communication aspects that involve adoptive processes, not least the message being communicated. It may be relevant to utilize a differentiated messaging strategy in the communication during the innovation-decision period.

From this study it appears that customer participation knowledge is important to early adopters and interaction knowledge is important to most majority adopters and to the laggards. The latter insight makes sense because of what we know of these adopter categories, especially laggards: they are generally suspicious of change agents, which may have to do with the fact that they do not know how to communicate with change agents. It, therefore, makes sense that getting knowledge on how to interact with staff in a store can affect the behavior of the laggards, making them more ready to adopt. The early adopters have a wider network than late majority adopters and laggards, and they may know who to ask. They just need to know what they need to do.

As has already been pointed out, any cause and effect relation in the adoptive process of an innovation is complex, and no simple cause and effect are likely to exist. While this study has indicated a cause and effect relationship between knowledge and rate of adoption, it is important to underline that there may have been many other factors involved in the adoptive process. The effect on the rate of adoption could, for instance, be ascribed to the influence of opinion leaders. They are typically early adopters ([1], p. 223), and it may be their buzzing that can explain the first increase in the rate of adoption identified in this study. Only further research can give us precise insight into this. However, this study can open up a discussion of new aspects of the innovation-decision period that need to be researched further. We may still be some way from completely understanding the diffusion process as it takes place in the twenty-first century, especially with respect to innovations in a digital age context. For instance, certain aspects of digitalization may be more difficult to grasp by some adopter categories.

This study has identified two types of process knowledge and two corresponding stages that can help nuance the innovation-decision process. However, while the findings of this study must be considered part of an ongoing research process to update diffusion of innovations theory, they may have value to practitioners until further research is carried out. Practitioners can avail themselves of communication models that focuses on the communication process, for instance, the AIDA model. This study has shown that it is relevant to use process knowledge when the communication messaging is related to the diffusion of innovations process.

One aim of this study was to test if current diffusion of innovations theory is still valid or needs to be updated or discarded, following the tradition of Charles Sander Peirce. The conclusion is that it appears that the diffusion of innovations theory needs to be updated. However, this study only relates to one aspect of diffusion of innovations theory, and what is represented here is only study. It is certainly relevant to ask if the present findings are generalizable before any findings become theory. One should always be careful to conclude any generalizability based on studies from just one country. Before answering any questions with respect to generalization, the study must be put in a cultural context. The cultural context is Denmark, a so-called Nordic welfare state, with an advanced economy. Denmark is a fairly small country, but this should not have any influence on the findings. Studies that have proven to have general validity have been carried out in much smaller communities than Denmark. There is no reason to believe that the adopter categories are very different across countries in the Western world, if they are exposed to the similar commercial and technological influences as those that are present in the country in which the study took place. Therefore, it is likely that the findings have validity in geographical areas that are similar or somewhat similar to Denmark, culturally, economically, and technologically.

Author details

Henrik Vejlgaard

Address all correspondence to: hev@cphbusiness.dk

Copenhagen Business Academy, Copenhagen, Denmark

References

[1] Rogers EM. Diffusion of Innovations. New York, NY: Free Press; 2003

[2] Fann KT. Peirce's Theory of Abduction. The Hague, Holland: Martinus Nijhoff; 1970

[3] Sriwannawit P, Sandström U. Large-scale bibliometric review of diffusion research. Scientometrics. 2015;**102**:1615-1645

[4] Degnan M. Recent work in Aristotle's logic. Philosophical Books. 1994;**35**(2):81-89

[5] Commens. Digital Companion to C.S. Peirce. 2017. Available from: http://www.commens.org/dictionary/term/abduction

[6] Popper K. The Logic of Scientific Discovery. London, England: Routledge; 2002

[7] Rogers EM. A prospective and retrospective look at the diffusion model. Journal of Health Communication. 2004;**9**:13-19

[8] Davis FD, Bagozzi RP, Warshaw PR. User acceptance of computer technology: A comparison of two theoretical models. Management Science. 1989;**35**:982-1003

[9] Baaren E, van de Wijngaert L, Huizer E. Understanding technology adoption through individual and context characteristics: The case of HDTV. Journal of Broadcasting & Electronic Media. 2011;**55**(1):72-89

[10] Atkins DJ, Neuendorf K, Jeffres LW. Predictors of audience interest in adopting digital television. Journal of Media Economics. 2003;**16**(3):159-173

[11] Dupagne M. Predictors of consumer digital television awareness in the United States. Communication Research Reports. 2006;**23**(2):119-128

[12] Chan-Olmsted SM, Chang B. Audience knowledge, perceptions, and factors affecting the adoption intent of terrestrial digital television. New Media & Society. 2006;**8**:773-800

[13] Hogg N, Lomicky CS, Hossain SA. Blog in the media conversation: A content analysis of the knowledge stage in the diffusion of an innovation. Web Journal of Mass Communication Research. December 12, 2008;**12**

[14] Moreau CP, Lehmann DR, Markman AB. Entrenched knowledge structures and consumer response to new products. Journal of Marketing Research. 2001;**38**:14-29

[15] Lee C, Coughlin JF. Older adults' adoption of technology: An integrated approach to identifying determinants and barriers. Journal of Product Management Management. 2015; **32**:747-759

[16] McCreadie C, Tinker A. The acceptability of assistive technology to older people. Ageing & Society. 2005;**25**(1):91-110

[17] Kratzer J, Lettl C, Franke N, Gloor PA. The social network position of lead users. Journal of Product Innovation Management. 2016;**33**:201-216

[18] Brocke, von J, Mueller O, Debortoli S. Class notes: Power of text-mining in BPM. BP Trends. July 5, 2016. www.bptrends.com

[19] Sepstrup P. Tilrettelæggelse af Informationskampagner. Copenhagen: Gyldendal Akademisk; 2010

[20] Parliament document. Paper File from the Information Office of the Danish Parliament. Copenhagen, Denmark; 2005. These are unpublished document data

[21] DTIO. Digital and Paper Files from the Media Office of the Department of Cultural Affairs. Copenhagen, Denmark; 2015

[22] Meyer G. Diffusion methodology: Time to innovate? Journal of Health Communication. 2004;**9**:59-69

[23] Vejlgaard H. Fast organizations: A comparative study of the rate of adoption in households and organizations. International Journal of Technology Diffusion. 2015;**6**(3):21-31

[24] Vejlgaard H. Late adopters can be fast: The case of digital television. Communications. 2016;**41**(1):87-98

Chapter 3

Project Communication Management in Industrial Enterprises (Step by Step)

Jana Samáková, Dagmar Babčanová,
Henrieta Hrablik Chovanová, Jana Mesárošová and
Jana Šujanová

Additional information is available at the end of the chapter

http://dx.doi.org/10.5772/intechopen.75160

Abstract

Communication is the basis of everything and is thus the key to effective project management. The question is "What is the relationship between the project management and communication?" Effective communication is one of the main elements of project management. Communication is a critical part between people, information, and ideas, and communication is the basis for project performance in organization. This chapter will be focused on "project communication management step by step." The chapter deals with all steps of project communication management which are defined as a combination of logical-related communication methods, tools, and techniques for a successful initialization, planning, implementation, control and administrative closure of the project communication. In the chapter communication environment (communication strategy, organizational structure), communication channel (communication methods, tools, frequency, and support of communication), communication cognitive (communication differences and skills), and communication system (feedback system and system of sharing and distribution of information) will be described.

Keywords: communication, project communication, project, communication environment, communication channel, communication cognitive, communication system

1. Introduction

Communication is an essential process in our day-to-day life, and the entire world revolves around it [1]. What is communication? Communication is everything. Communication is the sharing of information. Communication is the giving and receiving of messages. Communication is the transfer of information from one or more people to one or more other people [2]. Lasswell's maxim defines communication as "who says what to whom in what channel with what effect" [1]. The ability to communicate is an essential life skill and one that can be continually developed [3]. The word communication comes from the Latin word communis, which means common. When we communicate, we are trying to establish "commonness" with someone. That is, we are trying to share information, an idea, or an attitude among the team involved in that particular project [1]. Communication is a vital part of our daily routines. Students sit in school and listen to teachers. People read books and magazines. They talk to friends, watch television, and communicate over the Internet. The workplace is no different. Experts tell that 70–80% of working time is spent in some kind of communication [4]. Communicating is also at the heart of all good management. You cannot be an effective manager unless you know how to communicate confidently in whatever circumstances; this can be either on a one-to-one basis, with your team, or indeed if you are a senior manager or a director, to the entire organization [5].

On the base of these definitions, we can say that communication is the basis of everything and is thus the key to effective project. Even in biblical times, the importance of project communication was contained in the chronicle of the Tower of Babel, whereby it was reported that God caused a construction project to fail by interrupting communication through the creation of multiple languages [6].

What is the relationship between communication and project management? The answer to this question is not simple, because these two concepts are interrelated. Effective communication is one of the main elements of project management, and therefore it is necessary to be constantly engaged with communication [7]. Nowadays, the projects have large dimensions. They need a lot of time and resources to implement. They use a lot of funds, and it is unrealistic that each employee drove by itself. Therefore, project managers are responsible for projects. Project managers spend at least 80–90% of their time communicating on projects [8]. By the use of communication skills, project managers help to plan, direct, control, and coordinate their operations throughout the project life cycle. Most of the communication activities of project managers involve interpersonal communication and project communications, sharing information with project team members, and other stakeholders [9].

What is the relationship between the project management and communication? The answer to this question is not simple, because these concepts are interrelated. Effective communication is one of the main elements of project management [10]. Communication is a critical part between people, ideas, and information. Communication is the basis for project performance in any organization. And, therefore, it is necessary to be constantly engaged with communication.

The "Triple C model" talks about it. The importance of Triple C model can be summarized by the diagrammatic relationship: **communication** – cooperation – coordination = **project success**. Other elements of "C," such as collaboration, commitment, and correlation, are embedded in the Triple C structure. Of course, the constraints of time, cost, and performance must be overcome all along the way. The Triple C approach incorporates the qualitative (human) aspects of a project into overall project requirements. Triple C model facilitates better understanding and involvement based on foundational communication. The Triple C approach elucidates the integrated involvement of communication, cooperation, and coordination. Communication is the foundation for cooperation, which in turn is the foundation for coordination. Communication leads to cooperation, which leads to coordination, which leads to project harmony, which leads to project success [6].

If the project should be successful, a certain level of communication is needed that affects the success of the project. Underneath the level of communication, the **five main areas of project communication** can be seen in **Figure 1**: **communication environment** (communication strategy, organizational structure, project culture), **communication channel** (communication methods, communication tools, communication frequency, support of communication), **communication cognitive** (communication differences, communication skills), **communication system** (feedback system, system of sharing, and distribution of information), and **complementary part** (communication matrix and analysis of stakeholders).

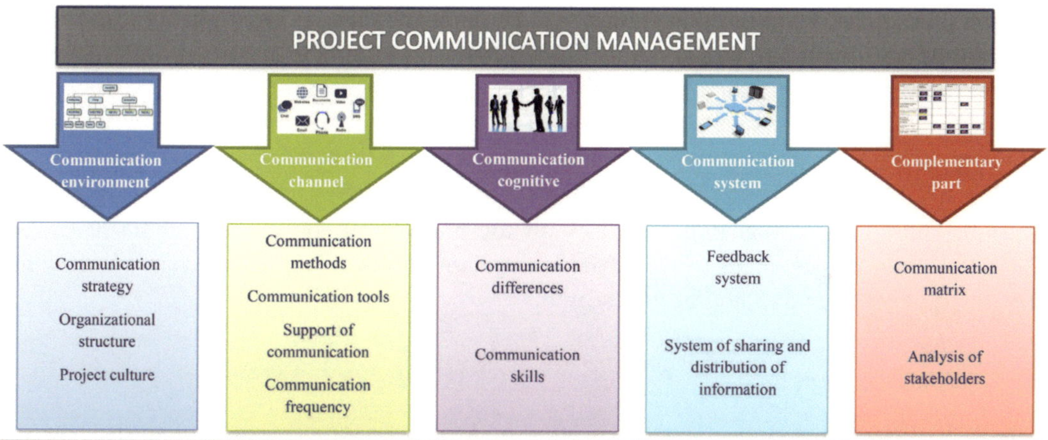

Figure 1. Five main areas of project communication management (own processing).

2. Methods and results

In this section of the chapter, theoretical research and empirical research of project communication management in industrial enterprises will be characterized.

2.1. Theoretical research

As the first point, we analyzed the project communication in international standards and methodologies of project management. For the comparison we have selected standard ICB® (IPMA® Competence Baseline) issued by IPMA® (International Project Management Association®), methodology PMBOK® (Project Management Body of Knowledge®) issued by PMI® (Project Management Institute®), and methodology PRINCE2® (Project in a Controlled Environment®) issued by OGC (Office of Government Commerce) in the UK.

ICB® describes the qualifications of project management. There are technical competences, behavioral competences, and contextual competences. Communication should be useful, clear, and timely. Communication under this standard may take many forms: oral, written, static or dynamic, formal or informal, and volunteered or requested [11]. PMBOK® is the most engaged one in project communication. Project communication management includes the processes required to ensure timely and appropriate generation, collection, dissemination, storage, and ultimate disposition of project information [12]. PRINCE2® is the least engaged one in project communication. The area of project communication is described in the following part "processes": The communication management strategy addresses both internal and external communications [13].

In the following section of the paper, we compare the basic international standards and methodologies of project management. Comparison of standardized project management methodologies is very difficult. Their different conception is very difficult for objective comparing. They are original, and they have widely varying vocabulary. They have different areas of knowledge, tools, techniques, procedures, material presentation, and other aspects of project communication. For comparison, they were used as the main areas of project communication, which are imaged in **Figure 2**: communication environment, communication channel, communication cognitive, communication system, and complementary part.

2.2. Empirical research

The analytical part of this paper focuses on the analysis of the current status of project communication management in industrial enterprises in Slovakia. The analysis was realized through qualitative and quantitative research.

2.2.1. Qualitative research

Qualitative research of project communication management served as a pilot study. The aim was to analyze how project managers understand the problems of project communication management in practice. In the research three medium-sized industrial enterprises (number of employees from 50 to 249) and one large industrial enterprise (over 250 employees) were interviewed and selected. The main results of qualitative research are imaged in **Figure 3**.

Basis formulated on the base of qualitative research:

- In the written documents the prescribed templates that make easier and clearer work in the project are absent.
- In the management of project communications, project managers frequently understand the creation of the "communication matrix" and treatment of the "stakeholder analysis."
- Project managers are interested in the project management methodology of communication, which will take into account communication environment, communication channel, communication cognitive, and communication system.
- For the management of project communication, project managers do not follow international project management methodologies and standards.

http://dx.doi.org/10.5772/intechopen.75160

Monitored elements	PROJECT MANAGEMENT METHODOLOGIES OR STANDARDS		
	ICB®	PMBoK®	PRINCE2®
COMMUNICATION ENVIRONMENT			
Communication strategy	☒	☒	✓
Organizational structure	☑	☑	☑
Project culture	☒	☒	☒
COMMUNICATION CHANNEL			
Communication methods	☑	✓	☑
Communication tools	☑	✓	☑
Support of communication	☒	☒	☒
Communication frequency	☑	☑	☑
COMMUNICATION COGNITIVE			
Communication differences	☒	☑	☒
Communication skills	☒	☒	☒
COMMUNICATION SYSTEM			
Feedback system	☑	☑	☑
System of sharing and distribution of information	☒	✓	☒
COMPLEMENTARY PART			
Communication matrix	☑	☑	☑
Analysis of stakeholders	☑	✓	✓

Explanatory Notes: ☒ methodology or standard does not include a specific element
☑ methodology or standard describes the element only briefly
✓ methodology or standard describes in detail, what the specific element addresses

Figure 2. Comparison of project communication in international methodologies and standards of project management (own processing).

2.2.2. Quantitative research

Qualitative research of project communication management was chosen because it is very good and easily quantifiable. Information about the management of project communication was surveyed by questionnaire. The research consisted of 128 industrial companies which use

PROJECT COMMUNICATION MANAGEMENT - general information	Enterprise A	Enterprise B	Enterprise A	Enterprise B
Size of enterprise	Large (over 250 employees)	Medium-sized (50-249 employees)	Medium-sized (50-249 employees)	Medium-sized (50-249 employees)
Sector	Automotive industrv	Engineering industry	Engineering industry	Chemical industry
Job position	Project manager	Project manager	Project manager	Project manager
Analysed project	Large project (minimum 12 months)	Large project (minimum 12 months)	Medium-sized project (minimum 3	Small project (minimum 2 months)
Methodology/standard of project management	Own methodology	Own methodology	PMBOK®	ICB®
Does your document – methodology of project communication management content following elements?				
COMMUNICATION ENVIRONMENT				
Communication strategy	✓	☒	☒	☒
Organizational structure	☒	✓	☒	☒
Project culture	☒	☒	☒	☒
COMMUNICATION CHANNEL				
Communication methods	✓	✓	☑	✓
Communication tools	✓	✓	☑	☒
Support of communication	☒	☒	☒	☒
Communication frequency	✓	✓	☑	✓
COMMUNICATION COGNITIVE				
Communication differences	✓	☒	☒	☒
Communication skills	☒	☒	☒	☒
COMMUNICATION SYSTEM				
Feedback system	☒	☒	☒	☒
System of sharing and distribution of information	☒	☒	☒	☒
COMPLEMENTARY PART				
Communication matrix	☒	✓	✓	✓
Analysis of stakeholders	✓	✓	☒	☒

Explanatory Notes: ☒ methodology or standard does not include a specific element
☑ methodology or standard describes the element only briefly
✓ methodology or standard describes in detail, what the specific element

Figure 3. Processing of qualitative research (own processing).

the project management. The survey was participated in 85 small, medium, and large industrial enterprises in Slovakia (128 respondents).

Basis formulated on the base of quantitative research:

- Industrial enterprises mostly do not have the written document, which deals with project communication.
- Managing of project communication can be considered as an important area within the project.
- Project management usually does not follow international project management standards and methodologies (ICB®, PMBOK®, PRINCE2®), but enterprises have increasingly developed its own methodology.
- Enterprises in the management of project communications take only a few into account elements such as multiculturalism, communication skills and communication strategy, and project culture.
- Industrial companies are interested in the developed project management communication methodology.

3. Results

Opportunities to improve communication within the project are numerous. The industrial enterprises can also create different communication models or different procedures for communication. In this contribution the scheme of **project communication management** is suggested. Project communication management is defined as a combination of logical-related communication methods, tools, and techniques for a successful initialization, planning, implementation, control, and administrative closure of the project communication (**Figure 4**).

3.1. Initialization of project communication

Initialization of project communication includes formulation of communication strategy and organizational structure of project communication.

3.1.1. Communication strategy

The goal of the **"communication strategy"** proposed by this contribution is to ensure effective sharing of information in the project. A well-processed communication strategy allows you to perform effective control of your work on the project. The communication strategy removes doubt, highlights planning, and includes all project stakeholders. The communication strategy should be elaborated by the project manager, together with the project promoter and the project team leader (if any). In the communication strategy, it is necessary to define:

(I) **Project environment**—the main goals of environmental research are to identify external opportunities and threats and identify strengths and weaknesses of the project. The study of the environment in terms of project communication is carried out through evaluation: the course of communication in past projects, the effectiveness of project communication in these projects, and perception of communication of stakeholders.

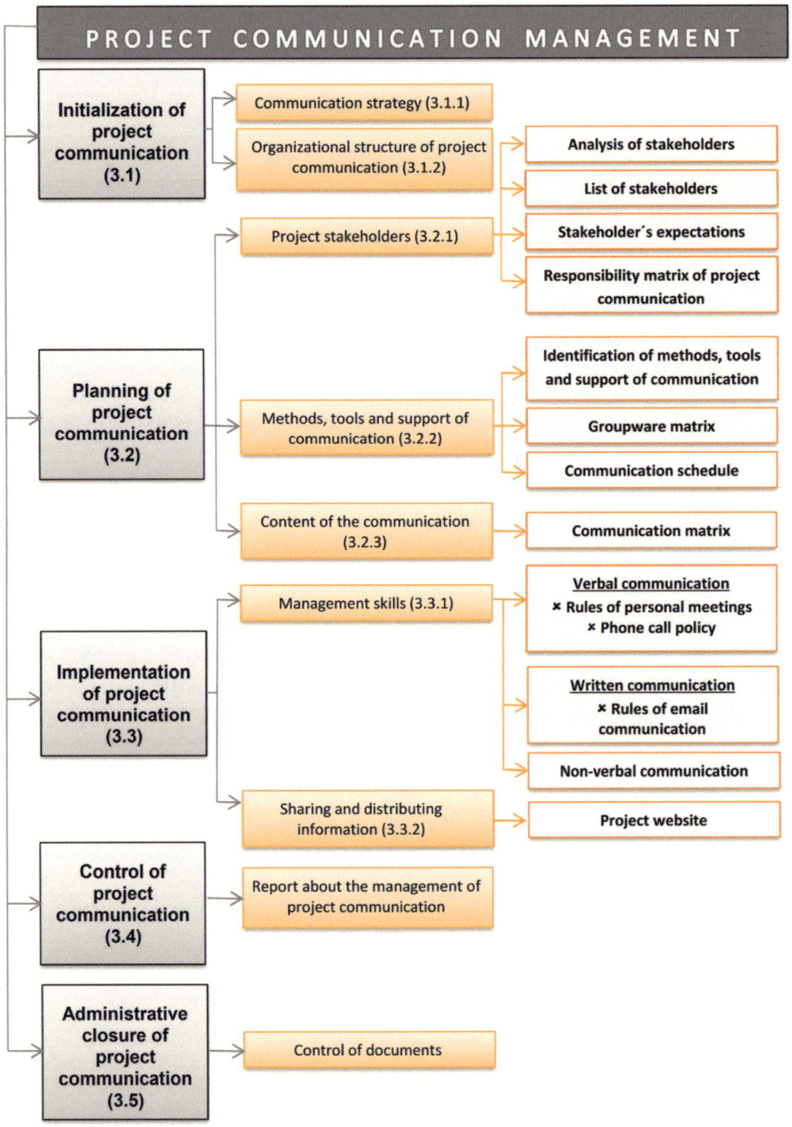

Figure 4. Scheme of the project communication management (own processing).

(II) **Project communication goal**—general objective (SMART: S, specific; M, measurable; A, assignable or acceptable; R, realistic or relevant; T, time-bound or trackable) and partial objectives of project communication.

(III) **Stakeholders involved in project communication**—it is very important to characterize all **internal stakeholders** (the executive director, the project manager, the project team leader, the project team, the quality department, the collaborators (e.g., the accountant, etc.)) and **external stakeholders** (key user (customer), key contractor, other stakeholders: environment (banks, offices, neighbors), experts (e.g., lawyer), etc.).

(IV) Communication channel—methods, tools, and communication support tools that will be used throughout the life cycle of a project.

(V) Type of information—the content of the information shared in the project is processed in the "communication matrix."

(VI) Principles and rules of project communication—contains project culture, stakeholders' requirements on communication, methods and tools, communication schedule, communication skills, control of project communication, and archiving of project communication.

(VII) Sources of project communication—within the framework of the strategy, it is necessary to define a project management budget and a system for sharing and distributing information in the project.

Regardless of the objectives of a project, defining a communication strategy shows attention to planning, an ability to carry out the work, and clear identification of the goal. The ability to communicate is essential to the success of any undertaking and an important factor in the achievement of its objectives. Communication does not just happen. It must be organized, developed, and built. A good communication strategy allows you to exercise better control over your work. A communication strategy removes doubt, emphasizes planning, and involves all the project participants [14].

3.1.2. Organizational structure of project communication

The goal of sub-process "project organization—**organizational structure of project communication**" is not designed as a new organizational structure of the project, but within the existing structures that industrial enterprises have established in their projects, they represent the organizational structure of project communication, in which the different types of communication are presented: formal communication, informal communication, and communication level (vertical, horizontal, and diagonal communication).

When managing a project, it is suitable to create a flat organizational structure that is much more flexible than hierarchical ones. The relationship between supervisor and subordinate is replaced by a wider range of communication, permitting for the use of new communication methods and tools within the project management.

3.2. Planning of project communication

Planning of project communication is an important process in the planning of the overall project. All stakeholders highlight about the importance and needs for project communication planning, but only 22% of them are concerned with these projects. The main reason for project communication planning is the success of the project.

Output of the planning of project communication process is the project communication management plan, which can be characterized as a tool for implementing the communication strategy of the project and answers the following questions:

- What information is needed for the realized project—the content of the communication
- When information is needed—frequency

- Who delivers information to anyone—responsibility and authority
- How information will be delivered—methods, tools, and support resources

Planning of project communication includes project stakeholders (analysis of stakeholders, list of stakeholders, stakeholder's expectations, responsibility matrix of project communication); methods, tools, and support of communication (identification of methods, tools, and support of communication, groupware matrix, and communication schedule); and content of communication (communication matrix).

3.2.1. Project stakeholders

In identifying stakeholders, it is necessary to define analysis of stakeholders, list of stakeholders, stakeholder's expectations, and responsibility matrix of project communication.

3.2.1.1. Analysis of stakeholders: the main project roles and responsibilities

In the project there are many groups of people involved in project management life cycles. The main project roles are project manager, project team member, steering committee, executive sponsor, customers, and stakeholders:

(I) Steering committee (project board)—steering committee is the highest appeal and decision-making body of the project. Its goal is to look at the progress of the project. The chairman is the executive director. The steering committee shall meet at the start and finish of the project and at the start and end of each milestone. In steering committee there are executive director, key user (main customer), and key contractor:

a. Executive director—it can be the project/program sponsor, referee, owner, executive sponsor, key user (main customer), and key contractor.

b. Key user (main customer)—key user represents the end users of the product or services from the project.

c. Key contractor—key contractor represents the suppliers (or subcontractors) for the project.

(II) Project **team**—project team is the group of people, who are responsible for planning and executing the projects. Project team consists of a project manager, project team leader, and a variable number of project team members:

a. Project manager—project manager is the person responsible for ensuring that the project team completes the project. A project manager is the person who has the overall responsibility for all the successful process groups: initiation, planning, design, execution, monitoring, controlling, and closure of a project.

b. Project team leader—project team leader is the person who leads the project team. Leader is responsible for contributing to overall project objectives and specific team deliverables. Project manager and project team leader can be the same person for small, medium, and large project.

c. Project team member—team members are people who actively work on the project, at same stage, during project management life cycle. Project team members are also responsible for contributing to overall project objectives and specific team deliverable. Team member may be able to cover multiple roles.

(III) **Project administrator**—project administrator or project coordinator is responsible for the maintenance of the project plan and maintenance and updating of a project web side.

(IV) Stakeholders—key stakeholders are individuals or organizations which are impacted by the outcomes of the projects.

3.2.1.2. List of stakeholders

The **"list of stakeholders"** is based on the organizational chart and is illustrated in **Figure 5**.

Project logo	**LIST OF STAKEHOLDERS**					
Project name and number:						
Function	**Name and surname**	**Phone**	**E-mail**	**Preferred method of communicati on**	**Preferred communication tools**	
					1.	**2.**
Statutory						
Executive director						
Key user – main customer						
Key contractor						
Project manager						
Project team leader						
Project team member (1 ....)						
Other stakeholders						

Drafted : (name + function): Date:.............................. Signature:..................................
Approved: (name + function): Date:.............................. Signature:..................................

Figure 5. List of stakeholders (own processing).

The list of stakeholders is as follows:

- Identification tool for all participants in the project
- Contact details secured by all project members
- Tool to ensure efficient and fast communication throughout the life cycle of the project
- Tool of mutually beneficial cooperation and communication between the project and its stakeholders

3.2.1.3. Stakeholder's expectations

After defining the roles and responsibilities of all stakeholders on the project and drawing up the list of stakeholders, it is necessary to develop a document named **"stakeholder's**

expectation" (**Figure 6**) within the project office management. The stakeholder's expectation document is a tool to identify stakeholders' expectations and their impact on the project.

Project logo	**STAKEHOLDER'S EXPECTATIONS**			
Project name and number:				
Function	**Name and surname**	**Stakeholder´s expectations**	**Impact on the project** supportive – S contradictory – C neutral – N	**Measures**
Drafted: (name + function):	Date:..............................		Signature:..................................	
Approved: (name + function):	Date:..............................		Signature:..................................	

Figure 6. Stakeholder's expectations (own processing).

3.2.1.4. Responsibility matrix of project communication

Individual roles, individual or group activities, and assigned responsibility must be redistributed among the parties involved. By incorporating these two elements, a **"responsibility matrix of project communication"** is created (**Figure 7**). Responsibility matrix of project communication displays the processes, threads, and their individual tasks in project communication. Developing the responsibility matrix of project communication is entrusted to the project manager, with the rows of the matrix showing the tasks and the columns of specific names of the persons (roles) responsible for the tasks.

Project logo	**COMMUNICATION MATRIX**					
Name and number of project:						
Role (Person) / **Task** (Process)	**Executive director**	**Key user – main customer**	**Key contractor**	**Project manager**	**Project team leader**	**Project team member 1**
Initialization of project communication						
Formulation of communication strategy						
Organization of the project						
......						
Planning of project communication						
......						
Drafted: (name + function):		Date:..............................		Signature:..................................		
Approved: (name + function):		Date:..............................		Signature:..................................		

Figure 7. Responsibility matrix of project communication (own processing).

3.2.2. Methods, tools, and support of communication

When planning project communication, it is necessary to define methods, tools, and support of communication, develop a groupware matrix, and process a communication schedule.

3.2.2.1. Identification of methods, tools, and support of communication

The **methods, tools, and support of communication** in **Figure 8** are arranged in order of their application in projects in industrial manufacturing enterprises.

Synchronous and asynchronous communication methods	Communication tools	Support of project communication
Synchronous – straight - Meeting - Personal interview - Phone call - Workshop - Conference - Social Activities **Synchronous - virtual** - E-conference *teleconferencing* *videoconferencing* *tele-videoconferencing* - Internet forum **Asynchronous** - Newsletter - Project documents - Letter - Board - Website	- E-mail - Telephone - Presentation - Video call - Fax - Paper - Unified communication - Chat - Internal chat - Social network - Video recording	- Microsoft Outlook - Microsoft Net Meeting - Calendar from company Google - Microsoft Office Communicator

Figure 8. Methods, tools, and support of communication (own processing).

3.2.2.2. Groupware matrix

"Groupware matrix" is used to capture and share information and knowledge, making work easier and faster. In case the information is to be sent and received in real time, synchronous groupware resources are used. In case real-time information is not required, asynchronous groupware resources are used. The matrix also uses distribution according to the segmentation of the team—receiving information in the same place (negotiation, meeting) or in different places (GARANT PARTNER PLUS, Všetečka—internal company material).

Groupware matrix:

- Is a tool facilitating multiparty communication and supporting collaboration, teamwork, and coordination
- Helps to work on common tasks
- Provides the information needed to perform the job tasks

3.2.2.3. Communication schedule

When managing and controlling a project, the project manager holds work meetings. Work meetings can be time-driven meetings and event-driven meeting [15]. The **"communication schedule"** shows all the meetings, workshops, conferences, newsletters, and more. Within the communication schedule, it is necessary to define start-up workshop, assistance workshop, kickoff meeting, planning workshop (risk management workshop, effort estimation workshop, cost calculation workshop, and others), work meeting, status meeting, project closeout meeting, regular meetings, progress meeting, emergency meeting, and so on. In addition to these appointments, sessions and meetings that are relevant to the effective functioning of the project are included: management committee meetings, project manager meetings, project team meetings, and external customer and supplier meetings.

Communication schedule is:

- A tool for identifying all planned communication methods
- A tool for planning individual meetings or sessions

3.2.3. Content of the communication: communication matrix

Communication matrix must be a part of every realized project. The matrix of communication is a written document, which is processed in the form of a transparent table and contains the following elements:

- ID: the number used to identify the content of the communication.
- Content of communication: the content of what needs to be communicated.
- Sender: information creator/owner responsible for sending the information.
- Recipient: to whom the information is intended.
- Methods of communication: way of communication.
- Communication tools: means of communication.
- Supporting means of communication: the selection of suitable means of communication.
- Frequency (periodicity) of communication: when (how often) the information needs to be communicated.
- Internal/external communication: this field indicates whether the communication is internal/external (I/E) or internal and external (I, and E) at the same time. The field is listed in the communication matrix to prevent the sending of messages to unauthorized persons.
- Comments: the way to capture responses to a message sent and delivered or the way to implement a message, e.g., in meetings it is possible to enter who is the facilitator of the meeting.

3.3. Implementation of project communication

The **implementation of project communication** is the third proposed process. Within the process, it is necessary to define communication skills and a system of sharing and distributing information.

3.3.1. Management skills: communication skills

In a part of implementation project communication process, rules of communication skills are designed which are necessary to be followed in project management to make the communication process as effective as possible.

3.3.1.1. Verbal communication

Within verbal communication, that is to a great extent affected by communication skills especially those of project manager, it is necessary to define the **rules of personal meetings**:

- Identifying one complex problem of the encounter
- Inviting to meeting only those people which are really needed
- Determining the time frame of the meeting to be followed
- To be interested in different views of all attendees (active listening)
- Use of a terminological dictionary
- Recapitulation of specific tasks arising from the meeting
- Building mutual positive relationships with all participants in the meeting

In verbal communication, it is necessary to deal not only with the rules of personal meetings but also with the **phone call policy**. These are the following rules:

- Determining who makes a short written record from a phone call
- Determining the person responsible for archiving the record
- Determining who gets written record from a phone call and who will distribute it
- Making a phone call written record

3.3.1.2. Written communication

In the context of written communication, it is important for the implemented project to establish the **rules of email communication**:

- Specify the persons who respond to the message (only the addressee or the persons mentioned in the copy answers).
- Determination of the persons to be indicated in a BCC copy and in a copy (CC) to BCC. The project address is given, e.g., projekt@projekt.sk.
- Specifying the priority of the message (e.g., answering important messages on the same day, responding to less important messages within 2 days, etc.).
- Determining who is responsible for archiving of the e-mail communications.

3.3.1.3. Nonverbal communication

Nonverbal communication can include, for example, visual contact, gesture, mimic, and so on. These areas are of major importance in every communication, but their greatest importance is growing in the implementation of multicultural projects.

3.3.2. Sharing and distributing information

The system of sharing and distributing of information serves to collect, store, process, and distribute information in the project. Project information must be relatively easily available to all participants in the project. A very efficient way is a **project website** that can be placed on a company website, or it can be a separate page. The website serves to share all the information about the project. The project manager will designate a person (internal or external) who will be responsible for the website of the project, whilst this person will be sending all outputs from the project in electronic form. The role of the project manager is to notify all project members about who is the person(s) responsible for the website, their e-mail addresses, and telephone contact. Project information will be stored on the page, such as organizational chart, project charter, stakeholder table, communication matrix, up-to-all meeting minutes, etc., so that all participants in the project have access to it.

3.4. Control of project communication

Control in project communication is very important. The aim of the project communication control is to develop the **report about the management of project communication**, which is part of a comprehensive project status report.

Report about the management of project communication includes the following particulars:

- The objective of control: whether project communication is taking place as proposed in the project communication management methodology
- Subject of control: verification of the efficiency of the processing of individual procedures and tools for project communication management
- The results of control: establishing corrective measures
- Identification data: the name of the implementer of the control, the processor of the report about the management of project communication, and the name of the authorizing officer who approve the report about the management of project communication

The principles of control of documents in a permanent organization are typically described in an internal directive in accordance with ISO 9001 standard. As every organization has the right to prepare its own standards of document control, it is necessary when planning a particular project to arrange an agreement with stakeholders on the adoption of standards. The rules of controlling documents are typically specified by the investor. In cases where it is appropriate and beneficial, the investor can decide on adopting the standards of another

http://dx.doi.org/10.5772/intechopen.75160

stakeholder, e.g., a supplier, if they have a well-prepared directive for the project management, etc. [15].

3.5. Administrative closure of project communication

The **administrative closure of project communication** is the final process. Administrative closure is important due to the creation of a large number of documents (written, spreadsheet, picture, audio, etc.) throughout the life cycle of the project.

For a purposeful processing of documents of the projects, it is necessary to draw up the document **control of documents**, which includes the following information:

- Document identification (assignment of registration number)
- Distribution of documents (sending of documents to project stakeholders)
- Document archiving (long-term storage of documents)
- Access and concealment of documents (identifying only those who have access to documents)
- Restoring documents (creating newer versions)
- Document destruction (destruction of unnecessary documents) (GARANT PARTNER PLUS, Všetečka—internal company material)

4. Conclusion

The world economy is changing with knowledge now being the primary value of an organization [16]. Enterprises perceive project management as a way to improve their competitiveness [17]. At the present time, many organizations do not want or do not consider important to train and to develop their own employees [18]. Communication is an integral part of any project, whether small- or medium-sized but especially large-sized projects. Communication influences most project activities and areas because managing any aspect of the project involves communicating within the project team or with external stakeholders [19]. Without effective communication, each project is convicted to be a failure; therefore, it is necessary to deal with project communication constantly. The most effective way is to manage project communication from the very start of the project—initialization of project until its completion and administrative closure of project. Project communication management is not an automated process because each project is different and needs different project communication managements. In small projects, project management is very simple and does not need to be planned and managed in detail. In medium-sized and large-sized projects, the number of team members and also the number of teams disproportionately grow over small-sized projects. Therefore, the project communication management is the necessary condition for the project. In **Figure 9** is shown, which part of project communication management are important during management of small-sized, medium-sized, or large-sized projects.

Project size	Small project	Medium-sized project	Large project
Main characteristics of the project:			
Number of team members	< 5	5 – 9	> 9
Number of teams in the project	1 – 2	2 – 3	> 3
Project duration	min. 2 months	min. 3 months	min. 12 months
PROJECT COMMUNICATION MANAGEMENT			
Initialization of project communication (3.1)			
Communication strategy (3.1.1)	☑	☑	☑
Organizational structure of project communication (3.1.2)	-	✓	☑
Planning of project communication (3.2)			
Project stakeholders (3.2.1)			
Analysis of stakeholders	✓	☑	☑
List of stakeholders	✓	☑	☑
Stakeholder´s expectations	✓	✓	☑
Responsibility matrix of project communication	-	✓	☑
Methods, tools and support of communication (3.2.2)			
Identification of methods, tools and support of communication	✓	✓	✓
Groupware matrix	-	✓	☑
Communication schedule	-	✓	☑
Content of the communication (3.2.3)			
Communication matrix	☑	☑	☑
Implementation of project communication (3.3)			
Management skills (3.3.1)			
Verbal communication			
- Rules of personal meetings	✓	✓	✓
- Phone call policy	-	✓	☑
Written communication			
- Rules of email communication	-	✓	☑
Non-verbal communication	-	✓	✓
Sharing and distributing information (3.3.2)			
Project website	-	✓	✓
Control of project communication (3.4)			
Report about the management of project communication	-	✓	☑
Administrative closure of project communication (3.5)			
Control of documents	☑	☑	☑

Explanatory Notes: ☑ compulsory; ✓ recommended; - optional

Figure 9. The use of individual parts of project communication during management of small-sized, medium-sized, and large-sized projects in industrial enterprises (own processing).

Author details

Jana Samáková*, Dagmar Babčanová, Henrieta Hrablik Chovanová, Jana Mesárošová and Jana Šujanová

*Address all correspondence to: jana.samakova@stuba.sk

Faculty of Materials Science and Technology in Trnava, Slovak University of Technology in Bratislava, Trnava, Slovak Republic

References

[1] Rajkumar S. Art of communication in project management. In: Project Management Institute, editor. PMI® Research Conference: Defining the Future of Project Management; 11–14 July 2010; Washington, DC, USA. Newtown Square, PA: Project Management Institute; 2010. 978-1935589136 193558913X

[2] Coates GT. Notes on Communication: A Few Thoughts about the Way We Interact with the People We Meet. 2009. 180 p. Free e-book from www.wanterfall.com

[3] Team FME. Effective Communications: Communication Skills. Free e-book from www.free-management-ebooks.com; 2013. 43 p. 978-1-62620-962-6

[4] Career Skills Library, editor. Communication Skills. 3rd ed. New York: Ferguson Publishing; 2009. 170 p. 978-0-8160-7778-6; 978-0-8160-7778-9

[5] Rowson P. Communicating With More Confidence. Hayling Island: Rowmark Limited; 2005. 156 p. 0 9539856 9 5

[6] Adedeji BB. Triple C Model of Project Management; Communication, Cooperation and Coordination. New York: CRC Press; 2008. 204 p. 978-1-4200-5113-1

[7] Svozilová A. Project Management. Praha: Grada Publishing; 2006. 356 p. 80-247-1501-5

[8] Kliem LR. Effective Communication for Project Management. New York: Auerbach Publications Taylor & Francis Group; 2008. 217 p. DOI: 978-1-4200-6246-5

[9] Verma VK. Human Resource Skills for the Project Manager: The Human Aspects of Project Management. 2nd ed. United States of America: Project Management Institute; 1996. 268 p. 1-88041-041-9; 978-1880410417

[10] Longman A. Management of people in project III. Communication, manager. Journal for Development of Senior Manager. 2007;**12**(47):28-30

[11] Caupin G, Knoepfu H. IPMA Competence Baseline. 3rd ed. Netherlands: IPMA International Project Management Association; 2008. 212 p. 0-9553213-0-1

[12] Project Management Institute. A Guide to the Project Management Body of Knowledge (PMBOK® Guide). 5th ed. Pennsylvania, USA: Project Management Institute, Inc.; 2013. 589 p. 978-1935589679

[13] Murray A. Managing Successful Projects with PRINCE2. 5th ed. London, United Kingdom: TSO—The Stationery Office; 2009. 327 p. 978-0-11-331059-3

[14] Geraldes J. Developing a Communications Strategy [Internet]. 2010-08-28 [Updated: 2018-02-07]. Available from: https://joaogeraldes.wordpress.com/2010/08/28/developing-a-communications-strategy/

[15] Všetečka P. A guide to the project thinking—Body of knowledge. PTBOKTM Guide. 1st ed. Liptovský Mikuláš, Slovak Republic: Peter Všetečka, Liptovský Mikuláš; 2015. 511 p. 978-80-9719-82-1-3

[16] Relich M et al. Measuring intellectual capital in the context of new product development. In: Cagáňová D, Čambál M, editors. Slovak University of Technology in Trnava. Proceedings of the 6th European Conference on Intellectual Capital; 10–11th April 2014; Trnava. Trnava: Academic Conferences and Publishing International Limited; 2014. pp. 153-160. 978-1-909507-20-3

[17] Brieniková J et al. The project management education in the Slovak Republic. In: Roman Kvasnička, editor. Efficiency and Responsibility in Education; 10-11th June 2010; Prague. Prague: Czech University of Life Sciences Prague; 2010. pp. 59-63. 978-80-213-2084-0

[18] Hitka M. Differences in employee motivation at Slovak primary schools in rural and urban areas. International Education Studies. 2015;**8**(5):33-42. DOI: 10.5539/ies.v8n5p33

[19] Muszyńska K. Patterns of communication management in project teams. In: Lecture Notes in Business Information Processing LNBIP 277, editor. Information Technology for Management: New Ideas and Real Solutions. 1st ed. Gdansk, Poland: Springer International Publishing; 2017. pp. 202-221. DOI: 10.1007/978-3-319-53076-5

Chapter 4

Aligning a Cybersecurity Strategy with Communication Management in Organizations

Ileana Hamburg and Kira Rosa Grosch

Additional information is available at the end of the chapter

http://dx.doi.org/10.5772/intechopen.75952

Abstract

Cyberattacks are a constant threat to organizations. Despite the improvement of cybersecurity (CS) techniques, criminals have developed sophisticated ways to disrupt systems and steal data particularly in organizations. The need to prepare them for cyberattacks is very important. **CS professionals** (CSPs), have a responsibility that should include protection against moral damage and protect moral rights to ensure the correctness, reliability, availability, and security of all aspects of information and information systems. In case of an emergency, effective communication is crucial. If IT systems fail, a quick communication with employees is necessary as well as to coordinate an effective response. This chapter starts with some explication about CS and later shows the problems organizations face when cyberattacks appear: The chapter describes methods for an efficient communication and its integration into CS strategy of the company. The scope of the chapter is to discuss with academics who work in the field of communication and CS and with students to find new scientific methods in this relative new domain based on some practical experience; and to help organizations and employees particularly CSPs to develop communication plans and to integrate them in their CS plans. The authors have experience will contribute that results will be integrated in the curriculum of cybersecurity from VET and HE students.

Keywords: cybersecurity CS, CS professionals, cyberattacks, cloud computing, platforms

1. Introduction

In the last years, cyberattacks are a constant threat to organizations. The companies and public offices have taken some cybersecurity precautions to strengthen security within the information technology field. The cybersecurity (CS) industry increases every year both the employment chances and the requirements at staff working in this field and at education. Despite the improvement of cybersecurity techniques, criminals have developed sophisticated ways to disrupt systems and steal data particularly in organizations. The need to prepare people and organizations for cyberattacks is very important. According to Cisco's 2017 Annual Cybersecurity Report (https://engage2demand.cisco.com/en-us-annual-cybersecurity-report-2017), more than one-third of the organizations that experienced a cyber breach in 2016 reported a loss of customers, business opportunities, and revenue.

In January 2013, cybersecurity strategy has been prepared by the European Commission to take precautions against the cyberattacks, which are performed, continuously to companies, public offices, and other strategically important offices.

The **CS professionals** (CSPs), who are individuals, which maintain CS, have a special role in preventing cyberattacks. They have a responsibility with a moral dimension that should include protection against moral damage and of moral rights to ensure the correctness, reliability, availability, and security of all aspects of information and information systems. In the event of a threat to security systems, the decisions of CSPs are very important. In case of an emergency, effective communication is crucial. If IT systems fail, a quick communication with employees is necessary as well as to coordinate an effective response. Two surveys sponsored by Websense and conducted by Ponemon Institute shows the damage that the lack of communication between CSPs, upper management, and employees can do in terms of overall performance of the company and public image.

There is no more literature about communication in case of a cyberattack being a recent and difficult topic. The authors presented besides their opinions the practical experience of consultants in the field of cybersecurity. The scope of the chapter is to help organizations and employees particularly CSPs to develop communication plans and to integrate them in their CS plans. Discussions with organizations within projects showed this necessity. The results will be integrated in curriculum of cybersecurity from VET and HE students who prepare CS field in the countries of partners of the project cybersecurity described shortly in this chapter. The topic of communication in CS is missing in the existing curriculum. The chapter is organized as follows:

In Part II of this chapter, some notions of CS and CS strategy are presented as well as the problems, which organizations have when cyberattacks came, and the role of CSPs.

In Part III, issues for planning a CS communication strategy are given, based on the experience of practical experts in this domain.

Part IV is dedicated to the methods for an efficient communication and its integration into CS strategy of the company. The scope is not to develop theoretical methods but to give practical help to the organizations in case of cyberattack and students who prepare in CS.

http://dx.doi.org/10.5772/intechopen.75952

An example of the current European project Cyber Security about CS with partners from education and industry from seven European countries is given in Part V. The project aims to develop measures to improve the training of future CSPs and CS knowledge of organizations in developing suitable CS strategies including communication as an important part.

The scope of this chapter is on the one hand to discuss with academics who work in the field of communication and CS and with students to find new scientific methods in this relative new domain based on some practical experience; on the other hand, the authors would like to help organizations and employees particularly CSPs to develop communication plans and to integrate them in their CS plans. The authors have experience in cloud computing and work in project about CS. It is planned that research and project results will be integrated in curriculum of cybersecurity from VET and HE students.

2. Cybersecurity

Information security (IS) and cybersecurity (CS) are very closely related terms and are used sometimes interchangeably. Richard Kissel gave the following definitions (https://www.quora.com/Whats-the-difference-between-cyber-security-and-information-security): Information security—IS is the protection of information and information systems from unauthorized access, use, disclosure, disruption, modification, or destruction in order to provide confidentiality, integrity, and availability. It is a broader field that is concerned with information and the protection of information whether be it physical or computerized [1].

Cybersecurity—CS is defined as the ability to protect or defend the use of cyberspace from cyberattacks. It deals with protection of cyberspace and use of it against any sort of crime. Confusion about terms is because most of the information today is saved electronically and most of the cyberattacks are executed to disclose confidential information, harm the integrity of it, or deny access to authorized users [2, 3, 4].

The information threats that do not involve cyberspace come under information security but *not* under cybersecurity.

CS is a broad term, which comprises the protection of critical information infrastructure from hackers, as well as elements, which are considered critical information infrastructures, such as information networks of small- and medium-sized enterprises or personal computers. CS strategy aims at preventing malicious cybernetic incidents, which affect both critical information infrastructures and noncritical information infrastructures. This has a purpose to protect goods and resources of organizations from the organizational, human, financial, and technical point of view, so as to allow them to continue their mission. Cyberattacks are a constant threat to organizations. Organizations must ensure that no significant prejudice is caused to them consisting in reducing probabilities that a threat materializes, in limiting the tried prejudice or deficiency and ensuring that, following a security incident, the normal functioning can be restored in an acceptable time frame and at a fair cost. CS is a complex process involving the entire society.

Referring to the design of a cybersecurity process and strategy, first it is important to correctly identify the goods and the resources, which must be protected, so that the scope of the security necessary for an efficient protection is precisely determined. This requires a global approach of security that must be multidisciplinary and comprehensive.

The elaboration of a cybersecurity strategy is necessary first due to the society's dependence on cyberspace, so that that security, resilience, and trust in information and communication field represent a problem of national interest. Secondly, economic role and possibilities of information and communication technologies and the intention to maximize benefits and exploit their opportunities are great [5].

Cybernetic attacks, especially committed against critical information infrastructure, could represent a threat to the national security and so cybersecurity strategies are related to security and national defense strategies. A cybersecurity strategy is necessary for the protection of confidentiality, integrity, and availability of data and information systems, to enhance security, resilience, authenticity, and trust in the field of information and communication technology [6, 7].

3. Communication issues and a cybersecurity communication plan

Almost in all countries, not only cybersecurity (CS) techniques have improved in organizations, but also criminals improved their ways to disrupt systems and steal data of persons and particularly in organizations. The need that organizations and employees are prepared for cyberattacks is very important.

In order to avoid decisions that could negatively affect organization reputation, a comprehensive and strategic crisis communications plan is necessary [8].

A communication plan requires a collective work, with the right roadmap and tasks not too daunting. One way recommended by Josh Merkin is to inspire from an old journalism trick and use the 5Ws: who, what, when, where, and why (http://www.odwyerpr.com/story/public/9215/2017-08-09/communications-plan-for-cybersecurity-breaches.html).

Why: The main objective of the plan is to prevent the loss of clients and revenue. Many firms do not have a crisis communications plan, i.e., for a data breach, often, due to less resources like time and money. Sometimes, the plan has no priority because it has no immediate or direct impact on business profit.

Who: First, it is important to establish who should be involved in developing the communication plan. A firm is managing partner/director, CEO, etc. A marketing or communication director is not probably the best decision; input should come from the firm's executive management team and IT (CS) department/team, legal counsel, administrative leadership, HR executives, and any communications agency and software vendors (if applicable). Strategic high-level input from senior leadership and department heads is necessary to ensure that all scenarios are covered (http://www.odwyerpr.com/story/public/9215/2017-08-09/communications-plan-for-cybersecurity-breaches.html).

The second important aspect is in case of a cyberattack; immediate decisions with potentially significant impacts will need to be made. So, activities outlined in the plan should be implemented quickly.

Within the planning process, it is important to determine who the key decision-makers are, how will they work when the time comes, and who is taking specific tasks.

In the third aspects, stakeholders including employees, clients, and possibly media, professional associations, law enforcement, and even government entities should be contacted.

What: The plan should include basic key messages and categories of information the firm will need to share with its audiences in case of a cyberattack. Information should be shortly adapted according to the situation, but a basic content should be put together in advance, including statement for press, internal and external memos, a news release, and messaging for the firm's digital channels and website.

Questions should be answered, i.e., if the attack was the result of an employee or software error, how much data was compromised and by whom? Some such scenarios could be written in advance, and it is beneficial to have all the key decision-makers involved in the development process *(*http://www.odwyerpr.com/story/public/9215/2017-08-09/communications-plan-for-cybersecurity-breaches.html*)*.

Where: It should be planned which channels the company has available to communicate its messages to its audiences, i.e., social media, email, phone trees, and directly to the person. For each case depending on situation and actions needed, it will be decided which channels are preferred over another or to use them all.

When: The point at which the plan will be implemented should be determined as well as how it will activate the response team should be activated. These issues should be specified in detail.

One particular role has the CSPs so it is expected that CSPs contribute to the development and application of a Telic communication plan for a worst-case scenario.

4. Issues to be considered by a communication/response

Multiple communication methods and channels can be affected in case of a cyberattack like own phone and voice mail system if they are VOIP-based, company phone system, company website (if it is hosted in-house), connections with customers, employees, the public, and the media *(www.continuitycentral.com/…communications…communications…/file).*

In case that the core network is compromised, every computer becomes a stand-alone machine with no access to company record. Employee contact information, vendor lists, or other key phone lists could be unreachable.

Besides a communication/response plan, some issues in connection with the IT infrastructure shall be regarded; affirm Nick Hawkins from EMEA.

Who needs to be included in an IT response plan?

- IT security responsible: should fix the issues and if the organization does not have a security team, employees must be assigned to be responsible for a response plan in case of a crisis.
- Incident team to coordinate the response, i.e., who should be contacted to define an escalation point.
- Legal counsel is necessary, i.e., if customer credit card details are stolen.

Who are the stakeholders?

Many stakeholders should be considered:

- C-Level executives
- Media relation department for messaging and informing customers about the incident and the press
- Customer services to prepare for incoming enquiries
- Employees to be kept up to date throughout the process to be prepared for calls from customers and the press
- Customers to be informed in real time about data breach

5. Cybersecurity internal communications best practices

In case of a cyberattack, the cooperative work and communication between the departments are very important in order to make sure that all security measures are aplite. The IT department should communicate with the chief information security officer (CISO)—who is a senior-level executive with duties including developing the company's information security architecture to best protect its systems and assets. In the following there are some true stories presented by Brad Berney (http://blog.securitymetrics.com/2015/03/internal-communication-for-it-security.html).

The head of the customer has to speak with the director of development when a customer found a security bug in their website (true story).

In case of an attack, the head has to action if the department is affected by lack of interdependent communication and customers whose data was stolen by a hacker due to such communication debacle.

In the following, some aspects will be presented to avoid communication problems within an organization. The first one refers to:

6. Cybersecurity communication culture

"It is possible that the poor communication culture from other companies gets thrown in the mix when employees are hired on from the outside. If an IT department hires three new employees from three very different technology companies, each will have a different expectation of how their team should operate" said Brad Berney (http://blog.securitymetrics.com/2015/03/internal-communication-for-it-security.html).

It is possible that industry branches in companies with the similar industry profile used different terms to describe it, and so some tension could appear.

Bad communication could determine loosing of skill employees than being demotivated.

Right communication involves also internal meeting where the employees can talk to each other about the cybersecurity problems. Sometimes, in the internal meeting, other goals have a priority, or the CSPs have no idea which is the right strategy.

Demotivated employees doing bad communication are at risk to leave the company. Brad Berney shows some such problems in a demotivating environment: "Nobody even cares about security around here (…) nobody even likes me in this company (…) nobody even asked me for that security report last month." (http://blog.securitymetrics.com/2015/03/internal-communication-for-it-security.html).

Brad Berney discussed with some unhappy CSPs and other employees and the firm that salary does not always play a role to motivate them for work, but they consider company culture and team communication as keys to success and happiness. A better and sure work environment is a good point to be happy.

Missing communication and company's diminishing security, particularly cybersecurity, are important factors often.

Berney told that on an audit he conducted, a company supervisor and he were confused why logs from the IDS/IPS are not being checked. When he asked, the IT employee simply stated "The alerts from the IDS were noisy, so I turned them off."

In the following we present some methods to improve communication in CS:

6.1. Improving communication in cybersecurity strategies through training

"Don't let employee training fall to the side of data security" said David Page, Security Analyst, QSA (https://de.search.com/web?q=microsoft+onlineportal&qo=serpSearchBox&qsrc=1).

Cybersecurity does not refer only to locks, firewalls, and the latest technology to protect employee's sensitive data but also their vulnerability.

Employees make mistake and hackers take advantage to access to data. Many cyberattacks and destroying of data happen because of unintentional employee actions which make organization business vulnerable i.e., by clicking a phishing email that downloads malware and gives sensitive information to someone or using non-protective passwords.

One common problem is that a cybersecurity strategy and security policies in an organization are requiring the employees who are not aware of them, i.e., to be informed about contained policy about on what to do if a cyberattack is supposed. In this case the employees could make an error or waste time in reporting it to the right people, potentially causing more damage for the organization [9].

Another problem is **social engineering**, which is rapidly becoming a big threat against businesses of all types and sizes. In security, social engineering is a broad term used to describe an information technology attack that relies heavily on human interaction and often involves tricking other people to break normal security procedures [10].

Social engineering refers to the techniques used to exploit human vulnerability to bypass security systems to gather information. Social engineering attacks imply interaction with other individuals, indicating also psychological and ethical aspects. About social engineering (SE), there are many differing opinions [11, 12].

Social engineering:

- Is known as human hacking
- Refers to the use of human error or weakness to gain access to any system despite the layers of defensive security controls that have been implemented via software or hardware
- Is the art of tricking employees and consumers into disclosing their credentials and then using them to gain access to networks or accounts

The problem with social engineering is that it targets employees specifically. If employees are not trained to recognize social engineering tactics, they could be vulnerable to a data breach. A moral of urgency should be trained in employees within CS strategies.

It is important to train employees on basic CS best practices, because problems like email phishing scans and social engineering can affect each employee in the organization. Employees with access to sensitive data should learn how to protect it.

Some topics for training could be:

(https://de.search.com/web?q=microsoft+onlineportal&qo=serpSearchBox&qsrc=1)

- Technology use
- Password management
- Data handling procedures
- Incident response plans
- Data security best practices
- Social engineering techniques

Regarding communication and meetings, it is not enough to hold yearly meetings because employees have to be aware to prioritize cybersecurity aspects in their daily activities. Some tips given by David Page are as follows:

(https://de.search.com/web?q=microsoft+onlineportal&qo=serpSearchBox&qsrc=1)

- Set monthly training meetings: focus each month on a different aspect of cybersecurity, such as passwords, social engineering, e-mail phishing, etc.

- Give frequent reminders: these could be sent out in an email or newsletter that includes tips for employees.
- Train employees on new policies ASAP: also, newly hired employees should be trained on policies as quickly as possible.
- Make training materials easily available: intranet sites are a great way to provide access to training and policy information
- Create incentives: reward employees for being proactive.

Through an active communication, all employees should understand that they have an important role in keeping business's data secure. Training of employees should be a top priority in each CS strategy.

6.2. Internal communication as a permanent task

Communication, also within CS, is a complex problem, and not each step can be defined before, but some aspects could be considered:

1. Department directors and CEOs should recognize a poor internal communication about CS.
2. Define training with some topics like:
3. The problem itself
4. How the problem is damaging company, employees, and customers
5. Clearly define process how communication should happen
6. What to do if feelings have been hurt
7. How complaints can be brought up
8. Hold interdepartmental "need" meetings focused on discussing what each department needs from the other, including timelines, milestones, and goals.
9. Address hurt feelings—Everyone has their own view on how certain issues should be handled.
10. Tell employees why—Sometimes employees just want to know "why" of things. Why are we buying this product? Why did not we buy the product I researched and suggested? Why did not we implement this solution? Why? When employees do not get answers to their "why's," they decide to make matters into their own hands. And, that's when security and process problems start. Remember, employees have the keys to the kingdom. You rarely hold anything other than the check book. Answer those employee questions as quickly and succinctly as possible.
11. Start fun communication exercises.

7. The role of CSPs

The **CS professionals** (CSPs) have a special role in preventing cyberattacks. In case of a threat to security systems, the decisions of information security professionals are very important. In case of an emergency, effective communication is crucial. If IT systems fail, a quick communication with employees is necessary as well as to coordinate an effective response. The survey, sponsored by Websense and conducted by Ponemon Institute, shows the damage that the lack of communication between CSPs and upper management can do in terms of overall performance of the company and public image (https://www.entrepreneur.com/article/235318).

Many CSPs believe that their organizations' security controls do not provide adequate protection against advanced cyberattacks, according to more than 5000 IT professionals from 15 countries including the USA. They affirm that executives do not put effective security controls in place and do not evaluate a data breach with financial loss. This is also the conclusion of a study conducted, also by the Ponemon Institute (https://www.ponemon.org/data-security), that the majority of CSPs professionals fail to communicate security risks effectively to upper management.

These reports show that along with managing and developing response plans against emerging security threats, cybersecurity professionals also need to inform upper management about the seriousness of security threats and convincing them to allocate adequate resources to protect against data breaches.

According to a study sponsored by HP Enterprise Security Products (http://www8.hp.com/us/en/hp-news/press-release.html?id=1571359), 30% of the cost of a data breach was due to business disruption or lost productivity. The study found that companies that invest in adequate resources develop communication plans, define a high-level security leader, and employ CSPs who have costs lower than companies that have not implemented these practices.

Some ideas are how the communication between CSPs and executives can be seen in the 2014 Websense-Ponemon report (https://de.search.com/web?q=microsoftonlineportal+login&qo=serpSearchBox&qsrc=1). The report found several key reasons why communication between executives and CSPs is so ineffective.

Security discussions occur at a low level and are rarely brought to executive's attention. Sometimes, CSPs warnings are too technical in nature and do not translate the threats into easy-to-understand language. Criticisms of existing practices are often filtered out before being presented to management.

Some helpful aspects:

- Ensure that cross-functional teams can communicate effectively and that awareness of these risks spread. People in engineering, sales, and marketing departments also need to be aware of security risks.
- CSPs must turn technical details of security risks into information that can be easily understood by upper management.

- CSPs should address these issues directly with the CEO and executive team broughing directly their attention and not be filtered out by intermediate players.
- As more data moves into the cloud and across other devices, companies face a greater risk of losing sensitive information to attackers or unauthorized users. According to Lobley (https://www.linkedin.com/in/colinlobley/), too many businesses fail to set quantitative parameters for risk (risk appetite) instead, to align the language.

In a real-world example, Lobley worked with a client outside the tech sector. When he asked frontline staff how the business was impacted from an incident that caused the IT system to go offline for 2 hours, the response was simply "not a lot." Upon talking to management, however, it soon became apparent that the company had exposed itself to significant risk.

7.1. Cloud computing offers many opportunities for communication platforms

An IT-oriented communication platform can be used for the following:

- Employee information: pushing information to employees about the company status and messaging
- Conference bridges: using Toll-free conference bridges for employee, vendor, senior management, board of directors, and other key stakeholder phone calls
- Stakeholder groups: using predefined groups that had been created for key stakeholders to push information via phone, text, or email

Cloud computing "is a model for enabling convenient, on-demand network access to a shared pool of configurable computing resources (e.g., networks, servers, storage, applications, services) that can be rapidly provisioned and released with minimal management effort or service provider interaction" (NIST definition). Cloud computing enables companies to use resources as a utility rather than having to build and maintain computing infrastructures in-house [13–15].

Most organizations trust on internal email to communicate in the event of a crisis, even though a cyberattack might impact the email network. In doing so, organizations are exacerbating the problem and potentially providing hackers with critical company information.

By having a system that operates entirely independent of an internal communications network, organizations can ensure that the bilateral lines of communication between management and staff remain open—even in the event of a cyberattack or IT outage that may compromise an internal network or a rush of calls which may overload a telecommunication network.

The benefits of selecting to use a cloud-based platform in the event of crisis are twofold. Firstly, they allow for location-mapping functions to be easily installed on employee's smartphones, meaning that business' can receive regular alerts and updates on their employee's last known locations. This wealth of data is then readily accessible should a crisis develops, ensuring that management is not only able to locate all of their staff but are also able to coordinate a more effective response, prioritizing and deploying resources to help those employees who are deemed to be at risk. Without this location-mapping function, businesses are being forced to rely solely on traditional routes of communication to find out if their staff are in.

Organizations with crisis management plans that include using a cloud-based location-mapping device are instantly able to know that Employee A is out of the impact zone and safe, while Employee B is at the epicenter of the crisis and likely to be in danger, making communication with them the top priority.

The second advantage to implementing secure, cloud-based communication platforms into a business' emergency communications plan is that it enables users to quickly and reliably send secure messages to all members of staff, individual employees, and specific target groups of people. These crisis notifications are sent out through multiple contact paths which include SMS messaging, emails, VOIP calls, voice-to-text alerts, app notifications, and many more. In fact, with cloud-based software installed on an employee's smartphone, there are more than 100 different contact paths that management can use to communicate and send secure messages to their workforce, wherever they may be in the world. This is a crucial area where cloud-based platforms have an advantage over other forms of crisis communication tools; unlike the SMS blasters of the past, emergency notifications are not only sent out across all available channels and contact paths but also continue to be sent out until the recipient acknowledges them.

8. The case of the European project cyber security

The European project Cyber Security (www.cybersecurityplus.org) with partners from education, research, and industry/business supports the European Cybersecurity Strategy.

The seven partner countries of the project are preoccupied to develop a strong strategy which is the sum of all national and international measures taken to protect the availability of information and communications technology and the integrity, authenticity, and confidentiality of data in cyberspace [16].

One difficult aspect is the preparation of future CSPs. Referring to cybersecurity education particularly in VET, the project partner countries like Germany do not have any body responsible for educational and professional training programs for raising awareness with the general public, promoting CS courses and communication in CS. There are no CS courses in vocational schools; this is a gap in the present, nor is a cybersecurity discipline included in the curricula of professional courses. So, one of the objectives of Erasmus + project Cybersecurity is to disseminate cybersecurity issues in formal and nonformal education and organizations, and fostering the development and skills of teachers, trainers, and CSPs will contribute to create a CS culture and communication strategies in organizations.

Through short research in European practices and education, development of a curriculum in cybersecurity education including communication strategies; organizing seminars and conferences in VET, HE, and other organizations; and development and distribution of a book about cybersecurity, including a chapter about communication the project will contribute in improving knowledge and skills of people in avoiding cyberattacks. The cooperation with the industry assures a practical character of the project outcomes [16, 17].

A workshop with academics, students, and representatives of the organization having the main topics (communication within cyberattacks) will be held this year in Gelsenkirchen, Germany.

A platform for communication and training will be also developed, and cloud computing will be used for a pilot environment.

9. Conclusions

The CS environment is rapidly changing. Cyberattacks are on the rise by using advanced technological means and are interesting because businesses use more technology. There are needs for rapid shifts in business strategies to adapt to other changes due to CS risks which change in scope and potential impact quickly. Organizations need to have the tools prepared to be able to communicate and recover quickly in the event of a crisis. The severity of a cyberattack and its impact depends on these factors.

Consequently, particularly CSPs must be prepared to communicate effectively in this challenging environment, using the best communication means and data for the right audience at the right time. The consequences of ineffective communication, resulting in misunderstanding security risks, can be catastrophic. Having a plan and understanding the business objectives, the stakeholders, their needs, and the risks themselves will help the CSPs to provide a clear, relevant message. It is important to ensure that communication is addressed to the right stakeholder group and then to verify that it has been understood.

The severity of a cyberattack and its impact depends on these factors. Critical communication platforms, a communication culture, corresponding training support in case of a breach to limit downtime and damage are important issues for further research of the authors. Students could learn how to develop an efficient, well-practiced incident response plan which can minimize cyberattacks damages.

Acknowledgements

The paper describes objectives and outputs of the European Erasmus+ project Cybersecurity.

Author details

Ileana Hamburg* and Kira Rosa Grosch

*Address all correspondence to: hamburg@iat.eu

Institute of Work and Technology, WH Gelsenkirchen, Gelsenkirchen, Germany

References

[1] Andress J. The Basics of Information Security: Understanding the Fundamentals of InfoSec in Theory and Practice: Elsevier; 2011

[2] Anonymous. FBI. Cyber-Attacks Surpassing Terrorism as Major Domestic Threat. https://www.rt.com/usa/fbi-cyber-attack-threat-739/

[3] Rattray G. Strategic Warfare in Cyberspace. Cambridge MA: MIT Press; 2001. DOI: 10.1006/tpbi.2001.1531

[4] Singer PW, Friedman A. Cybersecurity and Cyberwar. Oxford; Oxford University Press; 2014. DOI: 10.1111/cob.12074

[5] Engebretson P. The Basics of Hacking and Penetration Testing: Ethical Hacking and Penetration Testing Made Easy: Elsevier; 2011. DOI: 10.1016/j.phrs.2011.06.024

[6] Himanen P. The Hacker Ethic: A Radical Approach to the Philosophy of Business. New York: Random House; 2001. DOI: 10.1038/414933a

[7] Laurie GT. Genetic Privacy: A Challenge to Medico-Legal Norms. Cambridge UK: Cambridge University Press; 2002

[8] Davies J. Benefits of Cloud Communications in a Crisis". 2016. http://www.business-cloudnews.com/2016/06/09/benefits-of-cloud-communications-in-a-crisis-situation/

[9] Gulati R. The Threat of Social Engineering and Your Defense Against It. SANS Reading Room; 2003

[10] Maan PS, Sharma M. Social engineering: A partial technical attack. International Journal of Computer Science Issues. 2012;**9**(2):557-559

[11] Bisson D. 5 Social engineering attacks to watch out for. The state of security. security/security-awareness/5-social-engineering-attacks-to-watch-out-for/. DOI: 10.4158/EP-2018-0101

[12] Chitrey A, Singh D, Singh VA. Comprehensive study of social engineering based attacks in India to develop a conceptual model. International Journal of Information and Network Security. 2012;**2**(1):pp. 45-53

[13] Antonopoulos N, Lee G, editors. Cloud Computing. Principles, Systems and Applications. Springer International Publishing AG; 2017

[14] Assante D, Castro M, Hamburg I, Martin S. The use of cloud computing in SMEs. In: Procedia computer science 83, special issue: The 7th international conference on ambient systems, Networks and Technologies (ANT 2016) / The 6th International Conference on Sustainable Energy Information Technology (SEIT-2016) / Affiliated Workshops. 2016. pp. 1207-1212. DOI: 10.1007/s12350-017-0924-x

[15] Hamburg I. Improving e-learning in SMEs through cloud computing and scenarios. In: Gradinarova B, editor. E-Learning - Instructional Design, Organizational Strategy and Management. Rijeka: InTech; 2015. pp. 481-498

[16] Hamburg I, Grosch KR. Ethical aspects in cyber security. In: Archives of Business Research 5, no. 10, 2017. p. 199-206 PD, DOI: 10.1016/j.jaut.2017.12.006

[17] Warren MJ, Hutchinson W. Deception: A Tool and Curse for Security Management. IFIP/SEC 2001, 16th International Conference on Information Security. Paris, France, 2001. DOI: 10.1007/s00127-018-1529-7

Chapter 5

Neuro-Linguistic Programming and Managerial Communication

Miroslav Frankovský, Zuzana Birknerová,
Eva Benková and Ladislav Suhányi

Additional information is available at the end of the chapter

http://dx.doi.org/10.5772/intechopen.75301

Abstract

Neuro-linguistic programming (NLP) can be considered as a tool for the identification and change of communication behaviour. NLP is based on the concept of the construct of behaviour created by the series of stages, which are perceived as one action. In this chapter, the attention is paid to the NLP characteristics, NLP techniques, mainly to representational systems, rapport, pacing and leading. The chapter presents the results from the research of the assessment of NLP based on comparing the respondents who attended NLP trainings and those who did not attend NLP trainings and comparing managers and non-managers. The research results indicate the effectiveness of NLP in the context of managerial communication.

Keywords: communication, neuro-linguistic programming, managers, training

1. Introduction

The effectiveness of managerial work is multifactor-conditioned, and at the same time, it is examined within an interdisciplinary approach. This concept corresponds to the current conditions of an economic environment characterized by dynamics, turbulence, and continual changes [1]. The effectiveness of managerial work is related to the development of managerial competences, which enable to meet the presented conditions. Managers are expected not only to face the changes but also to suggest and implement these changes [2]. In this connection, an important role is also played by the form and way of managerial communication. From the point of view of the effectiveness of managerial work, mainly in the social context, the issue of managerial communication skills has a dominant position [3]. Effective managerial communication is one of the fundamental pillars of the activities for which the cooperation of people is inevitable. At the same time, it is one of the significant issues discussed in the context of managerial work [4], entrepreneurship, business [5], services, marketing [6] and many other areas of the economic environment.

The answer to the question how to communicate better, how to effectively communicate with our surrounding and how to communicate with ourselves is also offered by neuro-linguistic programming (NLP). It arose in the 70s of the last century, and its founders are Richard Bandler and John Grinder [7]. According to the authors, NLP is a set of models, abilities, and techniques for effective thinking, behaviour, and communication. Supporting results were also brought by the studies of Scott [8, 9]. Witkowski [10] pointed out the disunity of the research results that supported the NLP method or did not support it.

NLP is often presented as a magic tool for self-improvement of individuals and in this sense, it recently relies more on the assumptions of NLP rather than on qualitative and quantitative research. It is also caused by the fact that there are only few researches, which prove the reliability of NLP tools. In this context, we can mention the study of [11]. The author points out that the intelligent management by applying NLP in the techniques of interpersonal and intrapersonal communication in management can relate to the success of a whole organization. The given study was conducted in a cooperation of the Solidarity Union and large non-profit organizations in South Africa where five strategic leaders using and implementing NLP in their organizations were analyzed. The result is that NLP improves interpersonal and intrapersonal behaviour of the leaders, and this subsequently contributes to the success of organizations by using NLP in their behaviour, communication, and leadership.

It is obvious that the major part of examining the NLP method was conducted in the context of communication research [11]. NLP relates to the idea that the meaning of communication is in a reaction which it creates. Therefore, it is needed to adapt our communication until a required result, a reaction, occurs. In this context, according to [12], one of the first discoveries of NLP is the knowledge that experienced communicators use their language in order to create the climate of trustworthiness and understanding.

Determining the issue of NLP in the context of managerial communication assumes its specification from the point of view of conceptualization and operationalization. Within the conceptualization of the issue, the chapter presents individual concepts, approaches and opinions of several authors who pay their attention to the study in the area of neuro-linguistic programming.

2. Neuro-linguistic programming

Neuro-linguistic programming (NLP) is a model used by people to structuralize their unique life experiences. It is a set of abilities and techniques for effective thinking and behaviour [7]. Within NLP, several tools and techniques for discovering what represents perfection are presented [12]. NLP enhances the efficiency of the processes of remembering, forgetting, information evaluation and their displaying, perception, and responsiveness in all areas of a human being to a maximum support of a subjective and objective benefit, flexibility, and other abilities which are essential for satisfactory life-building [13]. NLP is an art and a science how to be excellent and it is also a study of subjective experiences. It helps to understand the differences between the things we do leading to average results and a failure, and the things we do which bring success and excellent results [14].

Knight [12] mentions that neuro-linguistic programming (NLP) is a process of modeling unique, own, conscious and unconscious patterns (of thinking, communication, and behaviour) used to continuously develop and strengthen our own potential. The author adds that modeling enables to reveal the uniqueness, specificity, quality, and the features of inborn talent as well as natural abilities of an individual. According to Knight [12], the essence of this approach lies in the study of perfection. It is an approach to learning, curiosity, and respect to a unique way of existing in the world.

According to O'Connor and Seymour [15], it is a set of models and skills for effective thinking and communication. Alder [14] adds that NLP deals with a whole structure of human experiences, and it tries to model the trains of thoughts, feelings and beliefs, the result of which is human behaviour. Hamill and Kerr [16] claim that the NLP concept shows that the major part of our seemingly autonomous behaviour is a constructed behaviour created by a series of successive stages which produce a final result. These stages run very often so quickly and spontaneously that we perceive them as one action. Hamill and Kerr [16] mention that NLP is based on the existence of programmed sequences of thinking and behaviour. It offers knowledge and tools for discovering the structures of these programmes and their effective influencing, change and using. NLP provides tools, techniques, and means for discovering the nature of excellent results in the way which can be coded and reproduced [12]. By accepting the fact that what we experience comes out from programmed sequences of thinking and behaviour, NLP provides us with the knowledge and tools for discovering the structures of these programmes.

In the literature, different NLP attributes are mentioned, for example, rapport, representational systems, modeling, metaprogrammes, definition of aims, anchoring, speech, body, asking questions in a right way, and many other techniques, which are connected to the setting of our mind, our values, attitudes and so on [12, 15, 17–22]. In this chapter, the attention is paid to representational systems, rapport and leading.

2.1. Neuro-linguistic programming techniques

According to O'Connor and Seymour [7], communication begins with thoughts, which we share with others using words, voice intonation and body language. When we are thinking about what we see, hear, and feel, we internally create these images, sounds, and feelings again. We again experience information in the sensual form in which we accepted it for the first time. Knight [12] claims that the way we think influences our actions. When we can understand how our thinking forms our actions, we can catch and repeat the moments of excellent behaviour.

2.1.1. Representational systems

There is a belief behind representational systems that every human being depicts the surrounding world, in other words, people represent the surrounding world in their brains preferably by one of the five sense systems [18]. Every person has one's own, the so-called, preferred sensual system which the person uses the most preferably and in which one feels the best. To depict experiences, we internally use the same neurological ways as by a direct experience. The same neurons generate an electrochemical charge which can be measured by an electromyographic reading device. A thought has direct physical effects, a mind and a body represent one system [7].

2.1.2. Rapport

Rapport creates an ideal state of communication based on trust and understanding, which can be used by managers and leaders when influencing behaviour, initiating a change, and persuading people [21]. According to Lazarus [19], rapport is one of the principles of success and can be defined as creating the spirit of trust and respect among people in order to create a higher probability of cooperation. Stone [23] claims that without rapport, there would be only a one-way communication. It requires practice and authenticity, and it means that we must be decided to accept opinions, observations, or motivation of other people. Bradbury [24] says that when we deal with another person, whether individually or in a group, we will always either follow or lead somebody. Regardless of how much we know about NLP, we will either:

- act similarly as the other person—in NLP, it is called following, that is pacing; or
- act absolutely differently than the other person—what is called leadership, that is leading.

2.1.3. Leading

Leading means leadership, accompanying, managing. According to Alder [25], leading means to have an influence on the behaviour of another person. Appropriate behaviour of another person is achieved by leading him/her when it is needed to become an agent of a change that is to create a change. This offers more control in communication. McCartney [26] states that adapting oneself to and creating a harmony with a partner for a certain time should lead to creating sufficient rapport, which subsequently allows to lead the partner where we want. By this way, we can use different representational systems of the partner. If the partner follows our leading, we have a good level of rapport. If the partner does not follow our leading, it is necessary to turn back to a harmony, to pacing and consequently to continue again with leading.

According to Alder [14], changes should be carried out gradually and not in a way which is too obvious. This technique can be used as a tool by persuading another person to think, feel and act differently. A therapeutic and pedagogical process, of any type, is always based on pacing, rapport and leading [13]. Basu [17] says that his understanding of a relationship is not only a basic principle of harmony and mirroring as understood within the NLP method.

In connection to the specification of the NLP issue, in the next part of the chapter, the attention is paid to the questions of operationalization of this issue.

2.2. Assessment of NLP

Based on the results of our research, it was possible to suggest and verify a new, original NLPT methodology—neuro-linguistic programming techniques. The managers were answering each item of NLPT methodology at a six-point scale of disagreement or agreement (0—definitely no, 1—no, 2—rather no, than yes, 3—rather yes than no, 4—yes, 5—definitely yes).

	Representational systems	Leading	Rapport
In communication with a partner, I use his style of expressing.	0.584		
In a conversation, I listen carefully to the words that a partner uses.	0.734		
In a conversation with a partner, I follow carefully his eyes.	0.733		
When presenting in front of people, I use several communication styles.	0.745		
I behave in the way which fits the other person the best.	0.616		
In a conversation with a partner, I try to understand his point of view.			0.588
To understand information properly, it is important to create a relationship with a partner.			0.550
What kind of relation I have with a given person is important when persuading this person.			0.723
Sharing information among people depends on mutual trust.			0.819
I communicate better with a person who has the same values, attitudes, and opinions as me.			0.581
By persuading another person, I try to change the thinking and behaviour of that person.		0.691	
When communicating to an annoyed person, I first get angry and after calming down, I get to what I want.		0.899	
Using emotions in communication makes my communication much easier.		0.588	
First, I try to change the opinion of my partner and then I expect the change of his behaviour.		0.803	
In communication, I use the experience of my partner which leads to his new way of thinking about a problem.		0.500	
Eigenvalues	3.466	2.716	2.628
Percentage of explained variance	23.105	18.106	17.519

Table 1. The extracted factors in the concept of neuro-linguistic programming techniques (NLPT).

Using a factor analysis in the concept of NLPT—neuro-linguistic programming techniques, three factors which specify three basic areas of perceiving the effect of neuro-linguistic programming were extracted (**Table 1**). Regarding their content, these factors can be characterized as follows:

- Representational systems—finding out and using a preferred method of thinking of a partner. The managers scoring high in this attribute use more often a partner's style of expressing when communicating. They perceive more carefully the way in which the partner thinks and uses words. They observe more the partner's non-verbal expressions. According to their need, they change and use several communication styles. (Cronbach's alpha 0.775).

- Leading—It is focused on the changes of partner's thinking and subsequent changes of his behaviour. The managers scoring high in this attribute try to use or better to say change the values, attitudes, and opinions of a partner in communication. They act so that they use linguistic means which their partner understands, and they proceed from his opinions. Subsequently, they offer their own solution (Cronbach's alpha 0.812).
- Rapport—an intimate relation between communication partners. The managers scoring high in this attribute try harder to create mutual trust and understanding of partner's attitudes. By persuading the partner, they focus more on the relationship with this person (Cronbach's alpha 0.781).

The extracted factors explain totally 58.7% of variance. Statistically highly significant values of correlation coefficients among the attributes of representational systems, leading, and rapport demonstrate the fact that these phenomena are closely interrelated, and it is inevitable to pay attention to all factors (**Table 2**). The given correlation indicates the fact that a higher degree of using a preferred way of partner's thinking is connected to a higher level of a mutual relation between communication partners and a higher level of leading.

The managers assessed the individual NLP techniques statistically significantly differently (**Table 3**). The managers positively assessed the importance of all studied NLP techniques. However, most positively, they assessed the importance of rapport as an intimate relation between the communication partners. On the contrary, the significance of leading, as focusing attention on the changes of partner's thinking and subsequent changes of his behaviour, was assessed negatively.

The extracted NLP attributes enable to describe and better understand the NLP construct from the point of view of NLP techniques. At the same time, from the methodological point of view, they offer a possibility to quantify the phenomena related to NLP.

2.3. Communication skills

Apart from the assessment of political, economic, social and technological contexts, effective solutions of the impacts of turbulences in the economic environment, which we mentioned in the introduction, also assumes a complex personality development of managers including communication skills [27]. Coping with the changes resulting from these turbulences requires from managers a high level of flexibility, creativity, and courage but also the ability to communicate effectively. Even when communication skills are included into soft skills, they significantly influence the success of managerial work. The importance of communication in managerial work is also emphasized by the fact that managers pay a significant part of their job to different forms of

	Leading	**Rapport**
Representational systems	0.448**	0.548**
Leading		0.437**

**Correlation is significant at the 0.01 level (2-tailed).

Table 2. The intercorrelations of the factors of NLPT—Neuro-linguistic programming techniques.

Factors	Mean	Standard deviation
Representational systems	3.451	0.7987
Leading	2.799	1.0375
Rapport	3.981	0.6997

Table 3. Assessment of individual NLP techniques.

communication [28]. Communication skills are therefore one of the priorities of selecting people onto managerial positions as well as of further education and the development of managers.

Effective managerial communication is conditioned not only by the expertise of managers but also by the way of communication. In the concept of communication, Annan-Prah [29] emphasizes that sharing information among its participants is not spontaneous but planned and organized with a precisely defined goal. The authors deal with the specification of communication skills and define many communication skills from different angles.

2.3.1. *Asking questions*

One of the important communication skills is the art of asking right questions in an appropriate way [30]. It is true that if we want to get a reasonable answer to a question, we must ask for it in a right way. By asking questions, we can obtain, specify and verify information, but also support or block individual communication. Therefore, asking questions must be thoughtful but also purposeful so that communication does not deviate from a chosen topic [31].

2.3.2. *Active listening*

In many concepts, we meet with the requirement of being able to listen to a partner in communication [28, 32–35]. The ability to listen to a partner in communication is not only one of the essential communication skills but also a primary condition for the effective communication. At the same time, active listening, by which it is outwardly clear that we are willing to listen to a partner in communication, creates an atmosphere of trust and interest in partner's problems, knowledge, information, and so on. Active listening also allows to verify the accuracy of the message interpretation by a communication partner. On the contrary, unwillingness to listen, the suppression of listening by one's own narrative, interrupting speech, verbal and non-verbal expressions of the interference reduce the effectiveness of communication, or they cancel it.

2.3.3. *Body language*

Identifying and interpreting non-verbal communication expressions is another important communication skill [25]. Body language, as many authors call non-verbal communication, includes the expressions of mime, haptics, proxema, posturology, gestures, kinesis, eye movements, paralinguistics and so on. Body language tells about the emotions and inner attitudes of a person, sometimes more than a spoken speech. It is the first expression, and it can reveal a lot to an experienced observer [36].

2.3.4. *Assertiveness*

Assertiveness is an important part and the condition of effective communications. Potts and Potts [37] state that the essential characteristic of assertive behaviour in the concept of profit-loss is the orientation towards the strategy of profit-profit. Lahnerová [38] claims that assertiveness is based on a natural human behaviour. We can adapt assertiveness to ourselves and our needs. The techniques of assertive communication include the DESC methodology [39]: D (description), E (expressing), S (specifying), C (stating consequences) as well as ERPG methodology [40]: E (empathetic understanding of a partner's position), R (respect of a partner's position), P (problem), G (goal). According to Frankovský et al. [41], the aim of assertive communication in the sense of sound self-assertiveness is the authentic expression of emotions and appropriate communication of one's own attitudes and demands and staying to one's own points. It is about acquiring adequate communication skills within interpersonal relationships, while respecting moral principles.

The four described communication skills of asking questions, active listening, body language and assertiveness were the subject of the research, the results of which are presented in the following sections.

2.4. Assessment of communication skills

It is clear from the presented research results that NLP relates closely to the issue of communication. From a methodological point of view, one of the possible approaches to examine NLP methodology is the comparison of the respondents who attended NLP trainings and the respondents who did not attend such trainings. NLP was studied in the context of communication skills based on the comparison of the answers of the respondents who attended NLP trainings and the ones who did not attend NLP trainings. The conducted research was focused on evaluating the impact of the knowledge from the NLP area on communication skills. The aim of the research was to confirm the impact of NLP trainings based on the verification of the existence of statistically significant differences in perceiving the communication skills among the respondents who attended or did not attend NLP trainings.

In the research, an original questionnaire CS—communication skills methodology, which contained 17 items, was used. The managers were answering each item of CS methodology at a six-point scale of disagreement or agreement (0-definitely no, 1-no, 2-rather no than yes, 3-rather yes than no, 4-yes, 5-definitely yes).

Based on the available literature and using a factor analysis in the concept of CS—communication skills methodology, four factors which specify four basic areas of perceiving communication skills were extracted (**Table 4**). Regarding their content, these factors can be characterized as follows:

- Asking questions—the skill to ask appropriate questions by which we can get important information (Cronbach's 0.870).
- Active listening—the skill to pay a focused attention to a communication partner (Cronbach's 0.802).
- Body language—it is the skill outside verbal communication (Cronbach's 0.763).
- Assertiveness—the skill to express our opinions, attitudes, feelings in communication with other people (Cronbach's 0.717).

	Asking questions	Active listening	Body language	Assertiveness
In conversation, I pay attention to the body language of my counterpart.			0.768	
There are situations when I consciously supress my own negative body language signals.			0.818	
Body language often says more about emotions and inner attitudes than verbal speech.			0.686	
Nonverbal accompaniment of my verbal speech follows the rule of the golden mean.			0.643	
When listening to my partner, I focus myself on verbal as well as nonverbal expressions of the partner's speech.		0.648		
When listening to a speaker, I show with my verbal and nonverbal expressions that I pay attention to the speaker.		0.586		
Listening is very important for conversation therefore I pay a considerable attention to it.		0.691		
My ability to listen is at a high level.		0.790		
I always try to achieve the satisfaction of both sides in communication.				0.409
If it is needed, I can say no in every situation.				0.602
If somebody behaves unfair to me, I would tell him/her about it.				0.756
I am open and decided by expressing my positive and negative feelings towards the other side.				0.727
In communication, I use one of the techniques of asking questions.	0.734			
By asking appropriate questions, I get precise information.	0.860			
I ask questions to have a clear idea what my communication partner thinks.	0.787			
I realize that by asking wrong questions, I can disrupt communication.	0.598			
My skill to ask questions is at a high level.	0.731			
Eigenvalues	6319	1990	1544	1104
Percentage of explained variance	37.170	11.704	9084	6495

Table 4. The extracted factors in the concept of CS—Communication skills methodology.

The presented factor structure is also supported by the values of Cronbach's alpha and the intercorrelations of factors (**Table 5**).

The managers assessed individual communication skills statistically significantly differently (**Table 6**). The managers positively assessed the importance of all studied communication skills. However, they most positively assessed the importance of active listening, the ability

	Active listening	Body language	Assertiveness
Asking questions	0.684**	0.390**	0.437**
Active listening		0.351**	0.470**
Body language			0.276**

**Correlation is significant at the 0.01 level (2-tailed).

Table 5. The intercorrelations of factors of the CS—Communication skills methodology.

to pay attention to a communication partner. On the contrary, the least positively (but not negatively), they assessed the significance of assertiveness, the skill to express our opinions, attitudes, and feelings in communication with other people.

It is clear from the presented research results that NLP relates closely to the issue of communication. From a methodological point of view, one of the possible approaches to examine the NLP methodology is the comparison of the managers who attended NLP trainings and the managers who did not attend such trainings. The concept of a research project presented in this chapter accepted the two mentioned attributes of examining the issue. NLP was studied in the context of communication skills based on the comparison of the answers of the managers who attended NLP trainings and the ones who did not attend any NLP trainings.

The conducted research was focused on evaluating the impact of the knowledge from the NLP area on communication skills. The aim of the research was to find out the effect of NLP trainings based on the verification of the existence of statistically significant differences in perceiving the communication skills among the managers who attended or did not attend NLP trainings.

In the research, we used an original questionnaire CS—communication skills methodology, described and characterized above. The questionnaire contained 17 items. The managers were considering each item of CS methodology at a six-point scale of disagreement or agreement (0—definitely no, 1—no, 2-rather no than yes, 3—rather yes than no, 4—yes, 5—definitely yes).

We compared the answers of the respondents who attended NLP trainings and the ones who did not attend any NLP trainings. A mathematic-statistical data analysis was carried out using the Student's T-test in the statistical programme SPSS 20.

Factors	Mean	Standard deviation
Asking questions	3.8774	0.88040
Active listening	3.9375	0.80182
Body language	3.6895	0.85026
Assertiveness	3.5544	0.72594

Table 6. Assessment of individual communication skills.

http://dx.doi.org/10.5772/intechopen.75301

	NLP training	Average	Standard deviation	T-test	Significance
Body language	Attended	3.49	0.812	2.858	0.005
	Not attended	3.92	0.843		
Listening	Attended	3.61	0.821	5.511	0.000
	Not attended	4.33	0.574		
Assertiveness	Not attended	3.31	0.789	4.382	0.000
	Attended	3.85	0.510		
Asking questions	Not attended	3.44	0.864	7.311	0.000
	Attended	4.41	0.543		

Table 7. Differences between the managers who attended NLP trainings and the managers who did not attend NLP trainings in perceiving communication skills from the point of view of extracted factors.

The discovered results were first presented from the point of view of the four mentioned factors of communication skills (**Table 7**), and subsequently, we analyzed the differences at the level of individual items which supplement these factors (**Tables 8–11**).

Table 7 presents statistically significant differences in assessing the factors of communication skills from the point of view of attending or not attending NLP trainings. We found out that the respondents who attended NLP trainings assessed the factors of communication skills more positively than those respondents who did not attend any NLP trainings.

The respondents who attended NLP trainings focus more on nonverbal communication using nonverbal expressions more often when communicating and listening more carefully to their partner in communication. Regarding body language, the respondents who attended NLP trainings confirm that they pay more attention to the body language of their partners in communication. They consider their ability to use nonverbal expressions to be highly developed, and their body language reflects their inner attitudes more than their

	Attended	Not attended	T-test	Significance
In conversation, I pay attention to the body language of my counterpart	4.41	3.60	4.989	0.000
There are situations when I consciously supress my own negative body language signals	3.22	3.22	0.030	0.976
Body language often says more about emotions and inner attitudes than verbal speech	4.40	3.90	3.210	0.002
Nonverbal accompaniment of my verbal speech follows the rule of the golden mean	3.76	3.16	3.632	0.000

Table 8. Differences between the managers who attended NLP trainings and the managers who did not attend NLP trainings in perceiving communication skills from the point of view of body language.

	Attended	Not attended	T-test	Significance
When listening to my partner, I focus myself on verbal as well as nonverbal displays of the partner's speech	4.13	3.62	3.105	0.002
When listening to a speaker, I show with my verbal and nonverbal expressions that I pay attention to the speaker	4.03	3.40	3.932	0.000
Listening for conversation is very important; therefore, I pay great attention to it	4.71	3.74	6.661	0.000
My ability to listen is at a high level	4.43	3.47	6.452	0.000

Table 9. Differences between the managers who attended NLP trainings and the managers who did not attend NLP trainings in perceiving communication skills from the point of view of active listening.

	Attended	Not attended	T-test	Significance
I always try to achieve satisfaction of both sides in communication	4.54	3.69	6.094	0.000
If it is needed, I can say no in every situation	3.68	3.03	3.774	0.000
If somebody behaves unfair to me, I would tell him/her about it	3.48	3.11	2.241	0.026
I am open and decided by expressing my positive and negative feelings towards the other side	3.60	3.34	1.949	0.050

Table 10. Differences between the managers who attended NLP trainings and the managers who did not attend NLP trainings in perceiving communication skills from the point of view of assertiveness.

	Attended	Not attended	T-test	Significance
In communication, I use one of the techniques of asking questions	4.22	2.74	7.341	0.000
By asking appropriate questions, I get precise information	4.51	3.45	7.057	0.000
I ask questions to have a clear idea what my communication partner thinks	4.52	3.59	6.816	0.000
I realize that by asking wrong questions, I can disrupt communication	4.56	3.97	4.785	0.000
My skill to ask questions is at a high level	4.40	3.17	7.797	0.000

Table 11. Differences between the managers who attended NLP trainings and the managers who did not attend NLP trainings in perceiving communication skills from the point of view of asking questions.

verbal expressions. Considering active listening, they pay attention to their communication partner using verbal and non-verbal expressions, and they are persuaded that their listening abilities are at a high level. With respect to the factor of asking questions, statistically significant differences were identified in all items which constitute the factor. Also in this case, the respondents who attended NLP trainings scored higher. The NLP trainings participants understand the importance of questions in communication, and therefore, they realize they must pay considerable attention to them. They apply the techniques of asking suitable questions in communication resulting in acquiring accurate information. They know that by asking inadequate questions, they can disturb a conversation and so they need to know what their partner in communication thinks. The differences in assertiveness between the respondents who attended NLP trainings and the ones who did not attend NLP trainings were recorded only in some indicators. In communication, they always want to achieve the satisfaction of both parties and if necessary, they say no in every situation. The biggest differences were identified in the items connected to the factor of asking questions. NLP represents a technique of unprejudiced questions enabling to put questions at a higher level or to approach the core of a problem.

The results of the analysis of the differences between the managers who attended NLP trainings and those who did not attend any NLP trainings assessing individual items, which supplement extracted factors also support the presented findings (**Tables 8–11**).

It is evident from the presented results that after attending NLP trainings, the managers assessed communication skills more positively not only at a level of the examined factors but also from the point of view of individual items, which describe particular aspects of communication skills. There is only one case related to the conscious suppression of one's own negative body language signals, from the point of view of which, no difference between the managers who attended and did not attend NLP trainings was found out. The given expression of communication skills is probably so significant that managers pay their attention to it regardless of having attended NLP trainings or not.

3. Discussion and conclusion

Many authors and users of NLP techniques have an opinion that NLP improves communication. Parkes [42] states that neuro-linguistic programming has become one of the most successful methods in the area of corporate as well as personal education. In this context, Singh [43] states that NLP offers managers and employees a significant improvement in the effectiveness of work, understanding and motivation of other people, individually but also in groups. According to Tripathi [44], by using NLP, we will understand our emotional and behavioral patterns of our behaviour. It is possible to apply NLP techniques to improve interpersonal, intrapersonal, and communicational skills of managers.

Neuro-linguistic programming is not only about understanding communication with the others but also about understanding oneself.

Understanding communication from the point of view of NLP provides us with the possibility to find the reasons how and based on what people communicate the way they communicate or they behave. The whole communication takes place in our heads, and it is therefore important to know that not circumstances but reactions to circumstances determine our behaviour in a given situation [45].

The results of the research speak in favor of confirming the effects of NLP techniques from the point of view of improving the communication with oneself as well as from the point of view of communication with other people [43]. They can be used, as already mentioned, in the context of managerial work [4], entrepreneurship, business [46, 47], services, marketing [5, 6] and many other areas of the economic environment [48, 49].

Based on the presented results and findings, we can think about the knowledge gained by attending NLP trainings as the knowledge which helps to improve communication skills. The presented results can be also used as a starting point for further research of NLP. From the point of view of further research, the question of the analysis of the connections between NLP and the development of communication competences of managers in the context of personal features of managers comes to the foreground.

Acknowledgements

This work was supported by the National Research Agency of Slovakia [VEGA 1/0909/2016].

Conflict of interest

The authors hereby declare that the research is original in its entirety and there is no conflict of interest.

Author details

Miroslav Frankovský, Zuzana Birknerová, Eva Benková* and Ladislav Suhányi

*Address all correspondence to: eva.benkova@unipo.sk

Faculty of Management, The University of Prešov in Prešov, Prešov, Slovakia

References

[1] Drucker PF. Management Challenges for the 21st Century. New York: Harper Business; 2001

[2] Peters TJ, Waterman RH. In Search of Excellence: Lessons from America's Best-Run Companies. New York: Harper & Row; 1982

[3] McKay M, Davis M, Fanning P. Messages: The Communication Skills Book. New Harbinger: Oakland. CA; 1995

[4] Daňková A, Droppa M. The impact of National Culture on working style of Slovak managers. Procedia Economics and Finance. 2015;**34**:164-171

[5] Bačík R, Mihal J, Fedorko R. The analysis of the impact of selected communication channels on the selected city population opinion. Polish Journal of Management Studies. 2015;**12**(2):7-14

[6] Korauš A, Štefko R, Dobrovič J. Decision-making satisfaction and behaviour of bank customers: Survey results analysis. Actual Problems of Economics. 2016;**186**(12):291-301

[7] O'Connor J, Seymour J. Úvod Do neurolingvistického programování [Introducing Neuro-Linguistic Programming]. Praha. Institut pro neuro-lingvistické programování; 1998

[8] O'Connell S. Harnessing power with your mind: Final 1 ed. Sunday Times. London: News International Trading Limited; 2006

[9] Albalawi KS. Effectiveness of neuro-linguistic programming on enhancing university students' quality of life. International Journal of Arts and Sciences. 2014;**7**(6):431-442

[10] Witkowski T. Thirty-five years of research on neuro-linguistic programming. NLP research data base. State of the art or pseudoscientific decoration? Polish Psychological Bulletin. 2010;**41**(2):58-66

[11] Oberholzer CH. The Role of Neuro Linguistic Programming in Improving Organisational Leadership through Intrapersonal Communication Development. South Africa: University of Pretoria; 2013

[12] Knight S. NLP at Work: The Essence of Excellence. London. Nicholas Brealey Publishing; 2010

[13] Lübeck W. Handbuch Des Spirituellen NLP. Windpferd: Verlag; 2003

[14] Alder H. NLP for Managers: How to Achieve Excellence at Work. United Kingdom: Piatkus Books; 2006

[15] O'Connor J, Seymour J. Introducing NLP Neuro-Linguistic Programming. Great Britain: Thorsons, (Reissue) edition; 1990

[16] Hamill D, Kerr D. An Introduction to NLP: Exploring What Works. What Works Consulting: Cheshire; 2013

[17] Basu R. Persuasion Skills Black Book: Practical NLP Language Patterns for Getting the Response you Want. UK Mainland: Bookshaker; 2009

[18] Fritzsche T. Jak přesvědčivě vyjednávat: Psychologické Strategie a Metody [How to Negotiate Convincingly: Psychological Strategies and Methods]. Praha: Grada Publishing; 2015

[19] Lazarus J. Part 2: Communication—Learning the many subtleties—Chapter 05: Getting people on your side: How to build trust with anyone. Richmond: Crimson Business; 2010. pp. 69-79

[20] Miller M. How to Persuade Anyone using Pacing and Leading [Internet]. 2017. Available from: http://www.mindwhirl.com/marketing/marketing-psychology/how-to-persuade-anyone-using-pacing-and-leading/ [Accessed: 2017-12-07]

[21] Perez C, Parker D. Creating Rapport to Influence. Enfield: Institute of Management Services; 2015

[22] Wake L. Applying NLP Tools and Techniques in an FMCG Environment. Hong Kong: Emerald Group Publishing, Limited; 2011. pp. 121-125

[23] Stone K. Influential People. Manager. Telford: Institute of Administrative Management; 2005

[24] Bradbury A. Manažerské Dovednosti: Jak ovlivňovat druhé [How to Influence Others]. Praha: Computer Press; 2001

[25] Alder H. Handbook of NLP: A Manual for Professional Communicators. New York: Gower Publishing Limited; 2002

[26] McCartney TK. The NLP Practitioner: A Practitioners Toolkit. United States: Lulu Publishing Services; 2014

[27] Kenton S. Studying Business and Management. Trotman and Company: Richmond; 2000

[28] Ferguson JG. Careers Skills Library: Communication Skills [Internet]. 2004. Available from: http://library.ikhzasag.edu.mn/ebooks/Communication%20Skills.pdf [Accessed: 2017-12-14]

[29] Annan-Prah EC. Basic Business and Administrative Communication. Bloomington: Xlibris; 2015

[30] Von Kanitz A, Scharlau C. Gesprächstechniken. Freiburg im Breisgau. Haufe Lexware GmbH & Co; 2015

[31] Knippen JT et al. Clarifying communication. Journal of Workplace Learning. 1999;**11**(5): 161-163

[32] Borg J. Talkability: Discover the Secrets of Effective Conversation. Pearson Business. Amazon Media EU S.à r.l.; 2016

[33] McLagan P, Krembs P. On-The-Level: Performance Communication that Works. Berrett-Koehler Publishers; 1998. 192 p

[34] Khelerová V. Komunikační a obchodní Dovednosti manažera [Communication and Business Skills of a Manager]. Praha: Grada Publishing; 2010

[35] DeVito JA. Essentials of Human Communication. Books a la carte edition. Cambridge, Pearson: Books a la Carte; 2013

[36] Bruno T, Adamczyk G. Körpersprache. München, Bayern, Deutschland: Rudolg Haufe Verlag GmbH & Co. KG.; 2009

[37] Potts C, Potts S. Asertivita—umění být silný v každé situaci [Assertiveness—The Art of Being Strong in Every Situation]. Praha: Grada Publishing; 2014

[38] Lahnerová D. Asertivita Pro manažéry [Assertiveness for Managers]. Praha: Grada Publishing; 2009

[39] Bower SA, Bower GH. Asserting Yourself: A Practical Guide for Positive Change. DA: Capo Press; 2004

[40] Stevens IG. You Can Choose to be Happy. "Rise Above" Anxiety, Anger and Depression with Research Evidence. California: Wheeler- Sutton Publishing Co.; 2010

[41] Frankovský M, Birknerová Z, Janovská A, Zibrínová Ľ. Vybrané kapitoly sociálnej psychológie pre manažérov [Selected Chapters of Social Psychology for Managers]. Prešov: Prešovská Univerzita V Prešove; 2012

[42] Parkes P. NLP for Project Managers. Make Things Happen with Neuro-linguistic Programming. London: BCS; 2011. 222 p

[43] Singh A. Neuro Linguistic Programming: A Key to Business Excellence. Abingdon: Total Quality Management & Business Excellence; 2008

[44] Tripathi S. Neuro-linguistic programming: A tool for developing behavioral skills and competencies. Hyderabad. IUP Journal of Soft Skills. 2012;**6**(1):16-28

[45] Krištofovič P. Minulé neúspechy Nie sú zárukou budúcich neúspechov [Past Failures Are Not a Guarantee of Future Failures] [Internet]. 2016. Available from: https://www.pkacademy.eu/sk/blog/minule-neuspechy-nie-su-zarukou-buducich-neuspechov/ [Accessed: 2018-01-08]

[46] Hudec O, Suhányi L, Urbančíková N. Regional decision-making criteria: Strategic investment in the Central Europe. Theoretical and Empirical Researches in Urban Management. 2014;**92**:104-117

[47] Suhányi L, Suhányiová A. Multi-criteria decision-making tool design for the investment decision-making of territorial self-government regions. Journal of Applied Economic Sciences. 2014;**91**:110-122

[48] Horváthová P. The application of talent management at human resource management in organization. In: 3rd International Conference on Information and Financial Engineering (ICIFE 2011); China: Shangai. 2011. pp. 50-54

[49] Mikušová M, Čopíková A. The identification of crisis manager skills by using Saaty's method. In: Proceedings of the 16th Eurasia Business and Economics Society Conference; Financial Environment and Business Development; Switzerland: Springer International Publishing. 2016. pp. 341-367

Chapter 6

Pharmaceutical Communication over Social Media Channels: 24/7 Management Challenges

Tiago Costa, Teresa Borges-Tiago and Flávio Tiago

Additional information is available at the end of the chapter

http://dx.doi.org/10.5772/intechopen.75990

Abstract

Social media marketing is evolving rapidly and gaining popularity within academia and firms. This chapter highlights the major challenges posed by social media to the pharmaceutical industry, assessing the presence of major pharmaceutical firms on social media platforms such as Facebook, Twitter, and YouTube and characterizing their digital engagement strategies. This study gathered data from the top 20 pharmaceutical companies' official websites, Facebook, Twitter, and YouTube accounts and examined variables such as Internet presence, engagement, fans/followers, and network structure. Findings from this study show that not all pharmaceutical firms are present on social media, and some platforms are used more than others. Notably, the digital engagement strategies varied between the analyzed social media platforms and remained similar on the two periods. Results also show that the level of engagement assessed was not associated with firm size. In several firms, the communication was mostly directed to the general public. Depending on the company, country-based communities were found. This work can be of interest to practitioners aiming to compare and assess their digital activity. It could also assist future researchers focusing on pharmaceutical social media marketing activity, since few researchers have analyzed this using more than one social networking site.

Keywords: communication, social media, pharmaceutical, Facebook, Twitter, YouTube

1. Introduction

Pharmaceutical firms are increasingly facing competitive, regulatory, and community pressures. For decades, pharmaceutical firms tried using different types of promotions to affect physicians' prescribing behavior. It has now evolved to a point where it has been often said that the leading firms were spending more on marketing than on research and development [1].

This led to misconceptions related to pharmaceutical marketing practices and induced the need to unveil the true meaning of pharmaceutical marketing. Pharmaceutical marketing, as a sub branch of marketing, evolved over the years from a more product-oriented strategy to a consumer-oriented strategy, following global trends [2].

Back in 1994, Levy [3] stated "pharmaceutical marketing is the last element of an information continuum, where research concepts are transformed into practical therapeutic tools and where information is progressively layered and made more useful to the health care system" (p. 327). His definition emphasizes the value of information flow. In the last two decades, there has been a major shift from traditional media to digital media, changing profoundly the way information flows. The race for information access turned patients, physicians, and all other healthcare stakeholders more demanding. In order to reduce these pressures and establish a long-term relationship with the different stakeholders, pharmaceutical firms need to follow marketing trends and move on to digital communication channels, such as social media.

The emerging importance of social media in business organizations is raising the awareness of decision-makers toward this theme. An ongoing dialog on platforms such as Facebook, Twitter, and YouTube between firms and consumers impels the search for new management communication models. In response, this chapter analyzes the online posture of 20 major pharmaceutical firms. For this purpose, data were gathered directly from their official social media pages on Facebook, Twitter, and YouTube, and their performance online was analyzed.

This chapter is organized in the following manner. The next section summarizes the literature related to pharmaceutical marketing and social media. The third section presents empirical evidence from the pharmaceutical firms. The last section offers concluding remarks about these findings.

2. Theoretical framework

2.1. Pharmaceutical marketing

According to Fischer [4], pharmaceutical marketing practice is mutating. The author states that the major approach in the 90s and the beginning of 2000 was increasing the sales force, allowing intensified diffusion of scientific information. However, Fischer explains that spending on this approach is decreasing over time. Nowadays, pharmaceutical firms are abandoning the commercial model based on sales force and are adopting a more multifaceted communication strategy. Through **Figure 1**, Fischer explains this new strategy, composed of more channels of communication which interface with other stakeholders (e.g., patients, payers, healthcare organizations, and professionals), surpassing the almost unique channel based on communication with physicians. The spreading of communication channels and potential message recipients is perceived as a major trend in pharmaceutical marketing [4] (**Figure 1**).

2.1.1. Physician-oriented model

Pharmaceutical marketing budgets are largely being allocated to communicate with physicians [4]. According to Fischer [4], in this type of communication, pharmaceutical firms

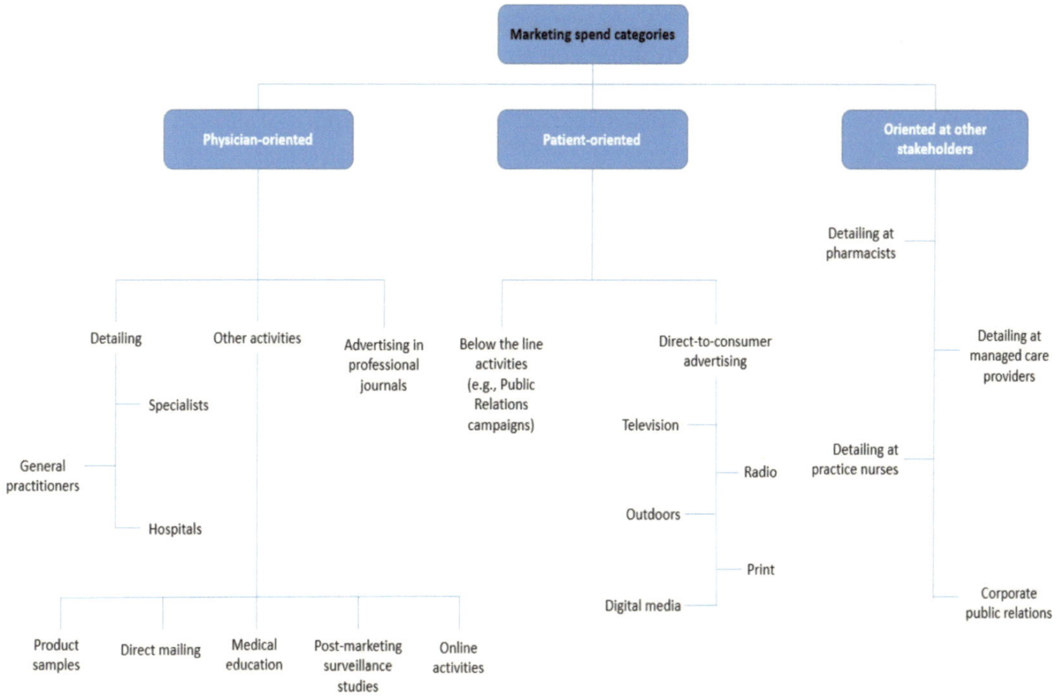

Figure 1. Pharmaceutical expenditure categories. Source: Adapted from [4], p. 560.

are using personal selling through detailing directed to physicians that are general practitioners, specialists, and hospital physicians. This regulation-dependent model can include (1) personal selling with discussion of a limited number of products, (2) delivery of some gifts and medication samples, (3) payment for meals, (4) conference-related issues, (5) financing of medical education and scientific projects, and (6) publicity in professional journals [4, 5].

Fischer states that methods like medical education and drug vigilance studies were developed to strengthen the relationship with physicians, and sometimes certain physicians can be provided with financial support from pharmaceutical firms if they are open to reporting their observations on patient trial studies associated with a promoted medication. Other channels, like direct mailing, allow pharmaceutical firms to deliver information about treatments or medicines to physicians [4].

Fischer [4] also considers that the traditional communication model is under fire from the rise of the Internet and other digital media. The author states that the study "Taking the Pulse® Europe" (p. 559), based on physicians' behavior on the Internet, revealed that physicians are using the Internet to diversify the reach of their influence [4]. Fischer refers to the findings in this study showing that in the main European countries, 95% of physicians queried said that the Internet is useful in their professional practice, and for this reason it seems that pharmaceutical firms are walking along with physicians in this matter through the development of electronic strategies.

According to several authors, physicians are open-minded toward technology adoption [6], allowing marketers to promote products through a new channel. It is clear that a growing number of physicians are networking online, and for that reason, pharmaceutical firms see the diffusion of information through online networking platforms as a priority [6, 7].

2.1.2. Patient/consumer-oriented model

There is a segment of pharmaceutical marketing that is being oriented toward patients/consumers, but this direct activity is only accepted in a small group of countries, like New Zealand and the United States of America (USA) [4, 6]. However, in the European Union (EU), pharmaceutical firms can advertise directly to consumers when the advertised product is an over-the-counter (OTC) medicine [8].

Fischer [4] points out that direct to consumer (DTC) advertising uses channels like traditional media (e.g. print media, television, and radio) and digital media. When DTC advertising is not allowed in a country, companies use a below-the-line strategy (e.g. Public Relations campaigns associated with diseases and treatments) in order to influence patients/consumers. Usually in this type of campaign, the branded medicine's name is not promoted [4]. Nevertheless, the global reach of online platforms poses a challenge in this diversified worldwide regulation setting [7, 9].

2.1.3. Model oriented at other stakeholders

Physicians and patients/consumers are not the only stakeholders targeted by pharmaceutical marketing [4]. Pharmacists, practice nurses, care providers, and informal caregivers are an audience that demands tailored communication strategies [4, 6]. Fischer [4] adds that the influence of these stakeholders as gatekeepers changes according to the type of medicine and healthcare structure. He provides an example with the healthcare business associated with diabetes. A major part of the budget is allocated to sales representatives in order to reach practice nurses and pharmacists since these healthcare professionals usually recommend devices that measure the blood glucose in people who suffer from this disease. Also, pharmaceutical firms allocate a percentage of their global budget into corporate public relations [4].

2.2. Pharmaceutical marketing challenges

Regulation is without a doubt a factor that distinguishes the pharmaceutical industry from other types of industries [10]. Regulatory activities are present in several pharmaceutical phases, like the approval of new medicines, medicine surveillance, production, and promotion to physicians and consumers [7]. The approval of medicinal commercialization normally requires a guarantee of safety and efficacy of the product. It is supervised by the Food and Drug Administration (FDA) in the USA, by the European Medicines Agency in the EU (even though individual country members have their own regulatory bodies), and by the Ministry of Health and Welfare in Japan [11].

To Desiraju and Tran [10], the regulation of marketing practices is not uniform, and several differences exist between different areas of the world and between pharmaceutical marketing segments. For example, since 2006 in the USA, it is not possible for a pharmaceutical company to communicate off-label medication uses to a physician, but the physician does not have restrictions on employing off-label uses for any situation in which they think the drug will work [7]. In other areas of the world, detailing practices are restricted, as well as medication samples given to physicians [7].

DTC advertising regulation differs from country to country, following the legal constraints of each country or region [10]. For instance, while in the USA DTC advertising of prescription medicines is authorized, in the EU and Canada there are restrictions for its use [10].

There have been some developments regarding DTC advertising in the EU, but they did not have a happy ending due to European Commission propositions being rejected by the European Parliament [12]. However, the European Commission, along with other industries that have interests related to medicine, is continuously pressuring the adoption of DTC advertising practices [12, 13].

In Canada, as DTC advertising is forbidden, pharmaceutical firms use two types of advertising allowed by regulatory institutions. The first type is associated with diseases, like "help-seeking" advertisements where non-branded communications increase awareness about a certain illness and appeal to consumers to seek their physician's guidance. The second type is a "reminder" advertisement that is a branded communication without any type of information about the use of the medicine [12](p. 635).

On the other hand, with the Internet, several concerns arise since online advertising can reach places where DTC advertisement is not allowed [12]. As an example, in the GlaxoSmithKline blog or the AstraZeneca Facebook page, the information is directed only toward the USA, but users from all over the globe can access this information, surpassing any possible control [12] (p. 635).

In terms of price regulation, there are also differences between nations [10]. It was also noted that price could be regulated through direct price impositions from governments, through price comparison between specific countries, or through pricing established by comparing medicines with an identical therapeutic category [7]. In some situations, governments can limit the global revenues of pharmaceutical companies. Countries like France, Italy, and Japan directly control prices, while countries like Germany, the Netherlands, and New Zealand manipulate reimbursements through price orientations, leading to changes in the amount that consumers have to pay [10].

Another challenge posed to this industry concerns the difference between branded medicines and generic medicines. A generic medicine is a trustworthy replica of a branded medicine, which has an expired patent, and is commercialized under the name of the active ingredient from the patented medicine [14]. With the rise of generic medicines, pharmaceutical firms stopped the nerve-racking approval process necessary for biosimilar branded medicines, allowing them to save time and money while diminishing process risk [5, 10, 14].

For pharmaceutical firms with branded medicines, the key target is physicians, while consumers and payers are vital targets for the development of brand loyalty and maintaining favored status in a formulary [5, 9]. These authors state that generic medicines are similar to commodities, and therefore, the marketing mix is focused on price. They also affirm that the key target audience of generic medicines is different from the target audience of branded medicines. Even though physicians choose medication for their patients, pharmacies are the last intervention in the supply chain of medicines to patients. For this reason, they can choose the companies that supply their stock, making them the ultimate targets for pharmaceutical marketers of generic medicines [5]. However, both physicians and pharmacies can receive financial support from payers in order to encourage the use of generic medicines [7].

2.3. Social media marketing objectives and outcomes

Social networks have been a reality for about 15 years, changing the way companies and consumers interact, both in the digital domain and by transposing into the physical realm. Both companies and researchers recognize that social networks have facilitated the narrowing of relations between companies and consumers, as they have provided consumers with an active voice and changed their behavior [15]. Therefore, the subject of social media is a high priority of discussion in the business world [16]. These authors argue that managers and marketers, as well as consultants, are forging new strategies aiming to increase companies' profitability through social media (e.g. Wikipedia, YouTube, Facebook, and Twitter).

Before social media, the Internet already allowed limited user interaction [17]. Web 1.0 (the first generation of the web) was considered to be the "read-only web." At this early stage, Internet users had limited interaction capability. Users were able to search and read information created and shared by firms [17]. Regardless of this interaction constraint, Web 1.0 revolutionized the access to information, since users could access countless diverse websites as information sources and were no longer limited to traditional vehicles like television transmissions or books [5]. Several of these websites appeared like simple brochures, whereas electronic commerce websites were similar to catalogs [5, 17].

The Web 2.0 was termed the "read-write web" [5, 17] since it allowed users to not only read but write content that could be shared to peers. This was the foundation of digital social interaction as we know it [17]. These authors recognize that with this iteration of the Web, online interaction was based on two-way communication. With Web 2.0, the sources of information considered relevant by consumers changed from firms to other online users [5]. These authors suggest that simultaneously with the rise of the Web 2.0, social network prototypes started to appear, providing new ways to facilitate interactions between people. These prototypes converted with time into modern social media platforms (e.g. Twitter and Facebook) [5].

To Kotler and Keller [18] "social media [is] a mean for consumers to share text, images, audio, and video information with each other and with companies and vice versa, encouraging brand engagement at a deeper and broader level than before" (p. 291).

Kietzmann, Hermkens, McCarthy, and Silvestre [19] contend that social media allows interaction between users based on seven functionalities: presence, sharing, conversation, identity, relationship, groups, and reputation.

From a user perspective, social networking sites (SNSs) integrate several digital formats of user-generated content (UGC) such as "blogs, virtual communities, wikis, social networks, collaborative tagging, and sites that allow shared media files" [20].

To most users, these types of SNSs are attractive because they enable them to participate in an ongoing consumption-creation cycle; the content created and shared by some become information sources for others, that can also be commented on, shared, and tagged [21].

From a brand perspective, the potential of social media lies not only in communication and sales, but also in persuasion [18]. Several researchers have pointed out the capability of electronic word of mouth (e-WoM) to influence peers' purchasing decisions [22]. Still, not all shared content has a positive impact on brands; as it happens in non-digital contexts, negative messages are also spread. Thus, social media can provide marketers a channel to be present on the web, reinforcing brand awareness and notoriety, but it also presents several challenges that need to be taken into account, some which depend on marketers' willingness to respond in an engaging manner to negative user comments [18].

Facebook presence has become a must for numerous companies, and Twitter can enhance business, even for small companies [18]. Social networks can leverage brand presence, brand awareness, and also reduce advertisement costs (especially important in fragile economic situations) [23]. However, it is important for companies to be aware of ethical matters (e.g. user privacy, spamming, publicity policies, data mining, and legal concerns).

2.4. Social media pharmaceutical marketing

Marketers are overwhelmed by performance promises related to new communications approaches and technologies. As noted by several authors, digital communication over social media with patients/consumers is becoming increasingly important in patient care and consumer decision-making [6, 24, 25].

Social networks allow the exchange of healthcare information associated with symptoms, possible diagnosis, treatments, adverse side effects, and medical evidence, as well as opinions about experiences with healthcare providers [26]. More, this information can lead to inappropriate decisions due to limitations in consumer health literacy [24].

In the healthcare system, there are two sides, namely patient side and provider side, that connect with each other through common platforms [6, 26]. The provider side congregates all agents related to the healthcare system, such as healthcare professionals, pharmaceutical firms, medical technology firms, and all managers and other professional groups that have access to patient data, and for this reason are able to influence the healthcare systems [6]. The patient-side integrates the patients, their informal support structure, and everyone searching for healthcare information or support. To interconnect the two sides, digital platforms must exist, such as physician-patient appointments, health services provided by hospitals and communities, organizations functioning as funding channels (e.g., Medicare), and other types of interactions between equipment/treatment providers and patients [26]. Some of these digital platforms support internal communication and information access, while others connect the healthcare providers and patients and are relevant in the gatekeeping and mediation processes, as well as providing information access [26].

Some social media functionalities are considered "Trojan horses" to the pharmaceutical industry. Social media networks such as Twitter can present only part of what needs to be known by patients, due to character limits for shared content, which may cause errors. With this type of content limitation, marketers are challenged to find the right types of content to capture and persuade users regarding a firm value proposition [16]. Adding to this, most users tend to consider SNSs as social and non-commercial environments, thus the presence of advertising content on these platforms could be viewed by users as inappropriate or offensive [18].

Another important aspect that needs to be taken into account is the fact the content is no longer solely created by firms. Thus, firms may lose control over the content and need to consider the electronic word-of-mouth (e-WoM) phenomenon.

The rise of e-communication through social media is changing the healthcare and pharmaceutical industry [27]. Patients and providers are being empowered, since social media allows a greater control over the creation of content. All these empowered players can establish interactive connections with the pharmaceutical industry using social media as a communicational channel, especially when pharmaceutical firms are considered trustworthy and present a value proposition that is perceived as positive by users [25]. Thus, social media platforms are acknowledged for providing wonderful opportunities for pharmaceutical marketers to interact with their customers and acquire knowledge of their markets [27, 28].

These social media platforms present a set of opportunities and threats for pharmaceutical marketers related to promotion and brand management. To better reach and engage users, they firstly need to learn from and follow up with users' voices in order to tailor their messages to each individual or group. Thus, the content created and shared needs to be customer oriented and optimized for different scenarios, taking into consideration technological limits and regulations, while promoting brand awareness or notoriety [5]. Pharmaceutical firms can gather important brand monitoring data through social media platforms and also obtain critical information about consumers [13, 27]. These two resources enable pharmaceutical firms to develop innovative marketing strategies and services that could increase brand awareness, improve customer loyalty, and increase users' trust and compliant behavior, while at the same time take advantage of the ability to promptly respond to information requests from different users. One of the major threats posed by social media regards privacy and security, since content needs to always preserve patient-physician confidentiality, according to regulations.

Patient-side social media sites are becoming important information repositories, used to acquire health data from sources other than healthcare professionals. Still, the motivation to use these communication channels is wider, including monitoring health professionals' decision-making, finding alternative treatments, predicting treatment results or consequences, and even self-diagnosis of minor symptoms [29]. There are several studies analyzing patient/consumer use of three specific SNSs: Facebook, Twitter, and YouTube (**Table 1**).

The fast growth of social media that facilitates online social behavior has significantly modified the nature of human activities, habitats, and interactions [23]. Patients become more

Reference	Unit of analysis			Main research areas
	Facebook	Twitter	YouTube	
[29]	x			Information-driven patient use of SNSs
[25]	x	x		Doctor-patient relationship
[24]			x	Content creation
[28]		x		Public health purposes
[30, 31]	x			Information source for patients living with chronic disease and their informal caregivers
[32]	x	x		
[33]			x	Education

Table 1. Some studies.

active and empowered by the seamless access to information. Social media provides an open channel to all healthcare stakeholders [6], with different countries' legal frameworks restraining or opening it, as a response to the public and policy concerns related to data privacy and security, as well as ethical behaviors.

Therefore, regardless of the main topic of research, the relevance of social media to the pharmaceutical industry is evident, especially to those that are clearly focused on the consumer. Thus, it is no longer enough to consider the impact and role of digital communication at the firm and industry level; a closer look at social media interactions is now mandatory.

3. Top 20 pharmaceutical companies: empirical evidence on social media

Since the emergence of social media, pharmaceutical marketing academics and marketers have been intrigued by the influence of such media on their field. In this chapter, we argue that some of the research gaps, especially those related to barriers as well as reasons to use social media, are slowly being filled in by academic research as the industry starts to use social media with a more active posture. Thus, the current chapter explores the following two research questions: How is social media marketing being used by pharmaceutical companies? And what's the influence of size and countries on social media activity?

In order to initiate this investigation, it was necessary to define the sample. As the subject of analysis was pharmaceutical companies' presence on social media, the first important step was to choose the size of the sample. It was decided to use the top 20 pharmaceutical firms within a ranking of the top 50 in terms of revenue, as elaborated by Pharmaceutical Executive in 2013.

After defining the size of the sample, the social media presence of the top 20 pharmaceutical firms (on Facebook, Twitter, and YouTube) was analyzed by visiting their global homepage in order to search for evidence of their activity on social media. Two research coders retrieved the data in two distinctive periods (May 2016 and May 2017). This month was chosen due to the fact that worldwide a higher search rate is presented for medications driven by season conditions (e.g., spring and autumn allergies). When the homepage did not have a reference to social media presence, other official websites of the companies (e.g., websites of companies in specific countries) were visited and analyzed.

Figure 2 suggests that not all the pharmaceutical firms are present in the three major social media networks.

Of the companies examined, only 12 (60%) are on all social media platforms analyzed, while merely two (10%), namely Takeda and Daiichi-Sankyo pharmaceutical companies, are completely absent on such platforms.

An interesting finding is that Twitter is the SNS chosen by most companies, even though it has communication constraints regarding content length.

With all data collected, the next step was to calculate the engagement on each platform for each pharmaceutical company, using the following formula, later used in a cluster analysis (**Table 2**):

The engagement level was assessed by applying the formulas present in **Table 3**—Engagement formulas and the results are presented below:

From the table's results, it can be inferred that Facebook engagement does not increase proportionally with a firm's revenue and number of employees, since Boehringer Ingelheim and Eli Lilly are the top performers in terms of engagement, in both periods. Similarly, Twitter engagement rates do not increase proportionally with the company size. However, one company (Teva) seems to have a high performance when compared to the other companies. The same result pattern was found on YouTube, where the size of the firm shows no correlation with the level of engagement achieved.

Looking at the activity of each SNS individually, starting with Facebook, two categories of posts were analyzed: (1) user posts and (2) brand posts. In terms of brand posts, all Facebook

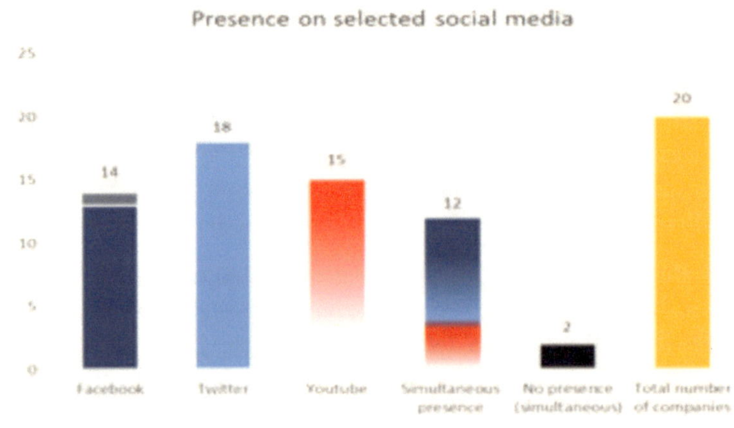

Figure 2. Pharmaceutical firms analyzed in terms of social media presence.

Platform	Engagement formulas
f	$\textit{Period engagement} = \dfrac{\textit{"People Talking About This'}}{\textit{nº of Fans}}$ $\textit{Period and Total engagement} = \dfrac{\textit{nº of Post} + \textit{nº of Shares} + \textit{nº of Likes} + \textit{nº of Comments}}{\textit{nº of Fans}}$
	$\textit{Period and Total engagement} = \dfrac{\textit{Retweets} + \textit{Mentions (including replies)}}{\textit{nº of Followers}}$
YouTube	$\textit{Total engagement} = \dfrac{\textit{Views per video}}{\textit{nº of Subscribers}}$

Table 2. Engagement formulas.

pages analyzed had brand posts on their wall. The company which had the highest activity was Boehringer Ingelheim with 73 posts while the two companies with the lowest activity were Roche (Genentech) and AstraZeneca, both with 10 posts. Regarding user posts, only six Facebook pages (Merck & Co., Roche, Abbott, Bayer, Boehringer Ingelheim, and Novo Nordisk) had user posts on their wall, with each one having its own differentiated activity (see **Figure 3**). Moreover, all Facebook pages registered comment activity. Furthermore, the company which registered the largest amount of comments was Boehringer Ingelheim, with 1660 comments, corresponding to 76% of all comments registered (in 2015). The figures for 2017 were quite similar, showing slights increases.

The second biggest Facebook page in terms of comments was Roche, with 208 (10% of the comments), demonstrating a large difference between Boehringer Ingelheim and the other companies. In order to have a more precise analysis of the comments within brand posts, it was necessary to display the comments per brand post. Boehringer Ingelheim continues in front with a ratio of 22.74 (50%) comments in 2015 and with 32.22 in 2017; Roche continues in second place, but a third company, GlaxoSmithKline, emerges passing from 4.69 (10%) comments in 2015 to 16.88 in 2017.

After analyzing the quantity of brand posts per company, a qualitative analysis of these posts was made. The majority of brand posts were in the format of the photo, totaling 59% posts in 2014 and 48% in 2017. In second place are posts with links, with 24% posts in 2014 and 32% in 2017. The least frequent form of posting was in a form of status posts and video in 2014. In 2017, video popularity increased, being used more (**Figures 4** and **5**).

This characterization reflects the performance of each cluster in each parameter, recognizing that these clusters are not homogeneous and therefore the companies included can have low and high performance in the same cluster (**Table 4**).

Ranking	Facebook	Score 2015	Score 2017	Twitter	Score 2015	Score 2017	Youtube	Score 2015	Score 2017
1	Boehringer Ingelheim	0.078	0.102	Teva	18.17	19.65	Abbott	49.103	47.425
2	Eli Lilly	0.074	0.083	Eli Lilly	9.971	10.362	Astellas Pharma	17.842	21.31
3	Roche	0.041	0.042	Merck & Co.	4.84	4.73	Bayer	10.905	15.006
4	Merck & Co.	0.038	0.034	Roche (Genentech)	4.238	4.547	Novo Nordisk	9.352	9.352
5	Roche (Genentech)	0.038	0.038	Johnson & Johnson	3.252	4.382	Roche (Genentech)	7.699	7.344
6	Teva	0.025	0.025	Novo Nordisk	2.425	3.126	Amgen	4097	6.524
7	Sanofi	0.015	0.019	Boehringer Ingelheim	2.122	4.702	Pfizer	2.88	2.99
8	Novartis	0.015	0.015	Gilead Sciences	1.782	1.885	AstraZeneca	2.595	3.198
9	GlaxoSmithKline	0.013	0.096	Bristol-Myers Squibb	1.603	1.715	Sanofi	2.163	2.192
10	Novo Nordisk	0.009	0.009	Astellas Pharma	1.257	2.055	Boehringer Ingelheim	2.088	3.702
11	Pfizer	0.007	0.01	Novartis	1.253	1.589	Eli Lilly	2.039	2.003
12	Abbott	0.007	0.007	GlaxoSmithKline	1.13	1.02	Johnson & Johnson	1.523	1.578
13	Bayer	0.006	0.008	Abbott	0.984	1.114	GlaxoSmithKline	1.437	1.474
14	AstraZeneca	0.005	0.007	Roche	0.83	1.013	Merck & Co.	1.138	1.235
15				AstraZeneca	0.824	0.992	Novartis	1.033	1.564
16				Bayer	0.817	0.977	Roche	0.917	1.246
17				Sanofi	0.73	0.72			
18				Pfizer	0.728	0.808			
19				Amgen	0.64	0.764			
20									

Table 3. Digital engagement and firm size.

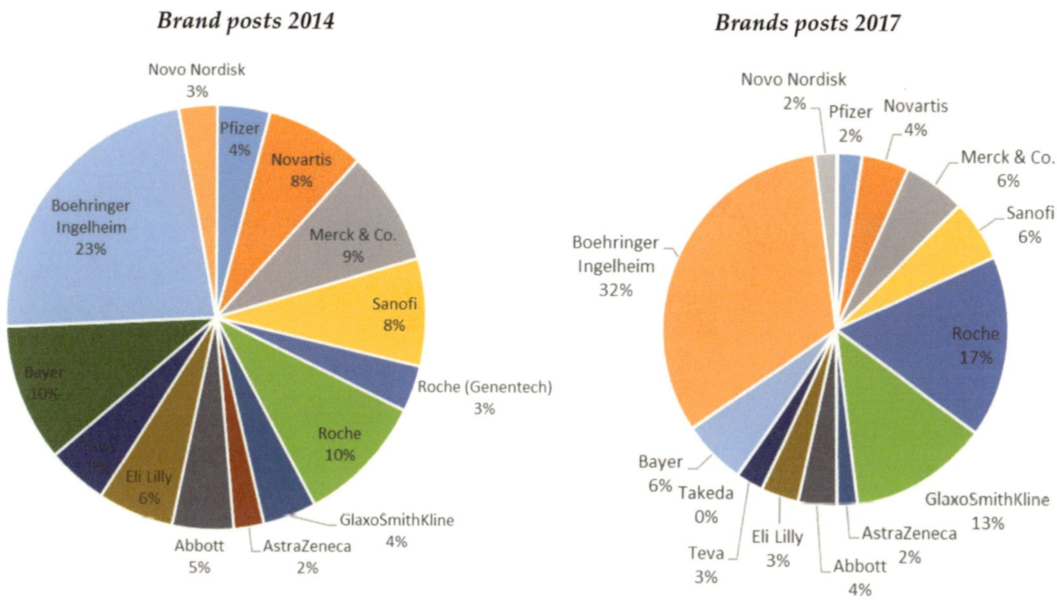

Figure 3. Activity on Facebook.

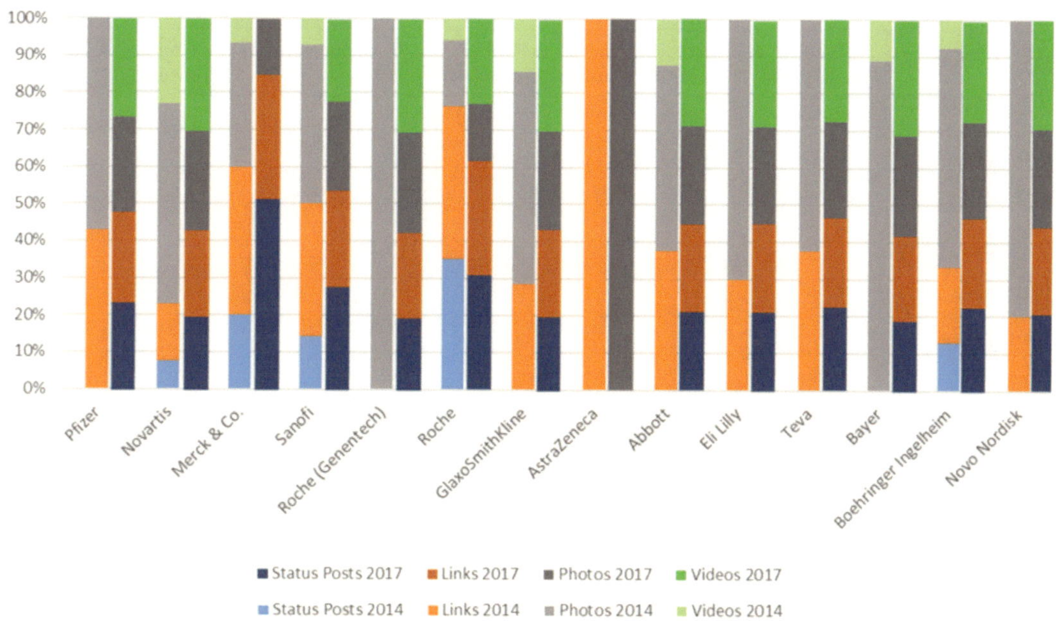

Figure 4. Type of contents posted by brands. The results from the cluster analysis conducted shows four distinctive clusters.

With regard to Twitter, 18 of the 20 pharmaceutical firms were present. Notice, however, that Teva's Twitter account is directed to the Spanish population, while the Astellas Pharma's Twitter account is dedicated to the USA population.

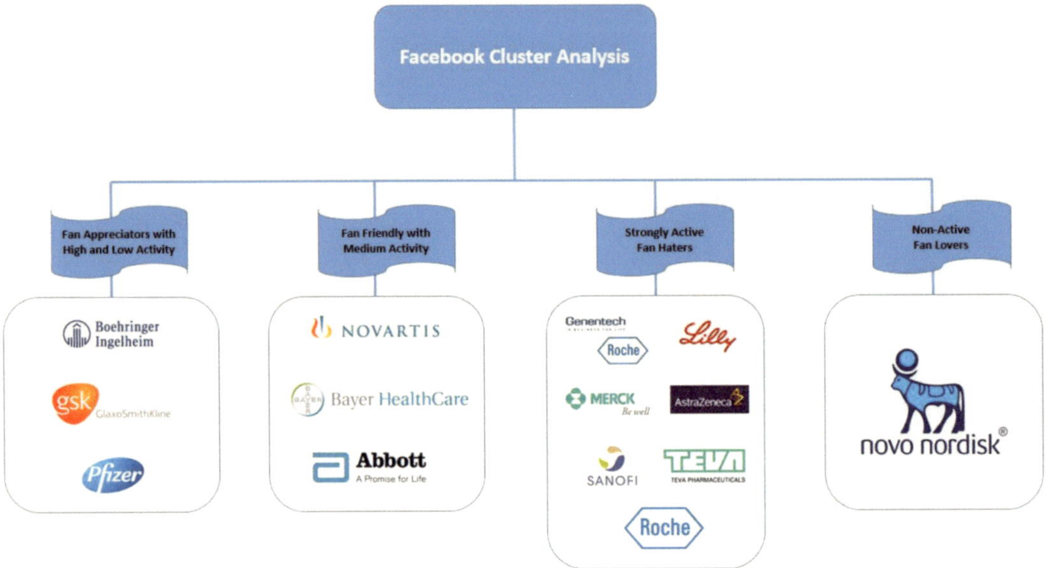

Figure 5. Facebook clustering agglomeration.

	Fan appreciators with high and low activity	Fan friendly with low to high activity	Strongly active fan haters	Non-active fan lovers
Revenue	+++/––	++/––	++/–	–
Employees	++/–	+++/–	++/––	–
Brand posts	+++/––	+/–	++/––	–
Brand post shares	+++	++/––	+/–––	+/–
Brand post likes	++	+/––	+/–––	++
Brand post comments	+++/–	++/–	++/––––	–
User posts	+/–––	++/–	+/–––	+/–

Note: Each + sign reveals the positive intensity in each parameter while each – sign reveals the negative intensity in each parameter, because clusters are not homogeneous.

Table 4. Facebook clusters: characterization.

In order to have a clear view of the activity on Twitter, three important aspects were analyzed: (1) tweets, (2) retweets, and (3) mentions, which includes replies. Three companies have high activity (over 5000 tweets) since their Twitter account were opened, with Boehringer Ingelheim leading with 6940 (16%) tweets, followed by Johnson & Johnson 6069 (14%) tweets and Novartis 5641 (13%) tweets. On the other hand, five companies registered less than 1000 lifetime tweets.

The customer service responses reflect the specific interaction between pharmaceutical firms and their twitter users' requests. The company most active in this category was Johnson & Johnson with 48 customer service responses, corresponding to 48% of their total responses. Interestingly, the only company to maintain responses in less than 1 hour was Boehringer Ingelheim during the period of analysis.

When analyzing the pharmaceutical companies' presence on Twitter through a cluster analysis, five clusters were identified (**Figure 6**).

After the first set of analysis, a deep analysis of the network structure of Twitter was conducted with the most recent data, using the NodeXL software, for the biggest and for the most engaging companies, aiming to establish their network structure and assess if it was linked to specific countries. The HK fast multiscale layout algorithm shows a pseudo-random network with five clusters in the case of Pfizer. Each cluster is linked to a specific country, with USA users the most active on the network (**Figure 7**).

Boehringer Ingelheim presents a quite uniquely structured network, denoting less spontaneous content creation and less community dispersion by country. The most relevant node is the company node. USA users are the most active on the network (31%), followed by UK users (16%), and Spanish users (9%).

The analysis of pharmaceutical presence on YouTube revealed that 15 companies have a YouTube Channel. The subscribers are an important part of the activity of YouTube channels because they receive notifications for each new video. One remarkable standout in the subscriber analysis is the dominance of Johnson & Johnson, registering the highest number of subscribers, while the second biggest company is Amgen. Johnson & Johnson outperforms in this category with 58% of the total number of views registered. Pharmaceutical companies'

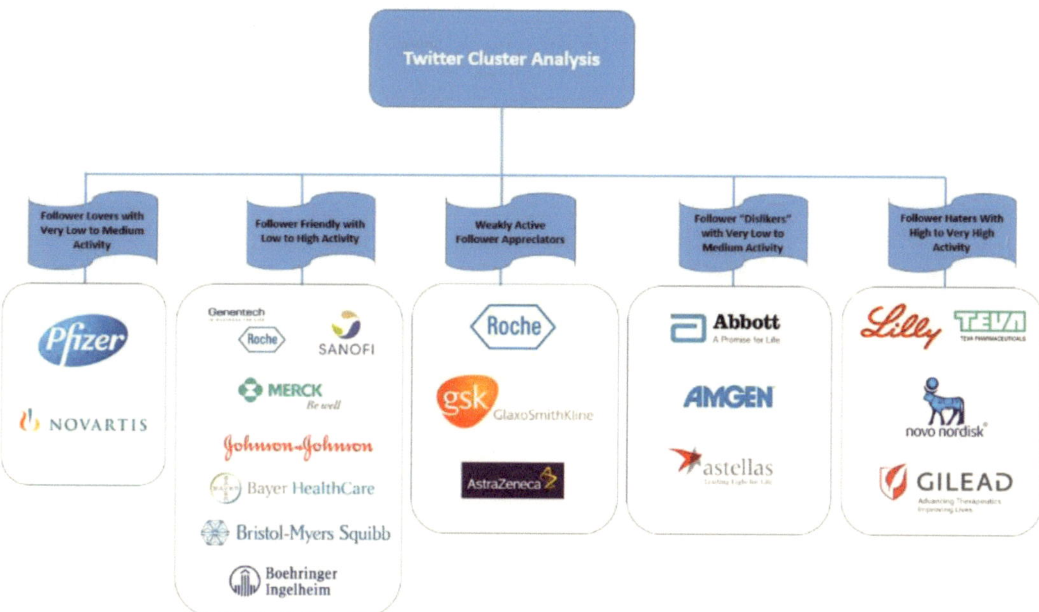

Figure 6. Twitter cluster analysis.

YouTube channels were also subjected to a cluster analysis identical to the ones performed for Facebook and Twitter (**Figure 8**).

The Japanese companies Takeda and Daiichi-Sankyo have completely abandoned the three assessed platforms, while Astellas is not in the same situation because Astellas USA possesses a Twitter account and a YouTube channel. YouTube appears to be the most abandoned social media platform, even though the use of video contents increased.

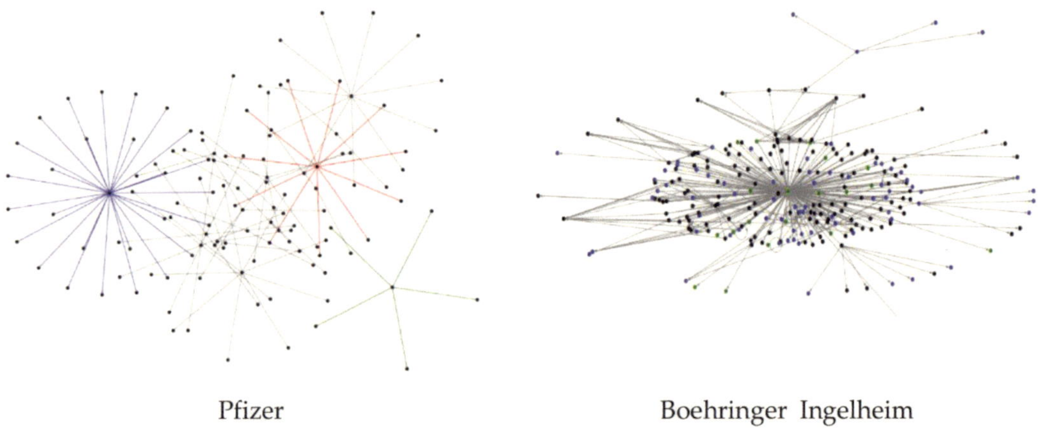

Figure 7. Network graphs obtained using the Harel-Koren Fast Multiscale layout. Legend: blue—USA, red—Greece, green—Germany.

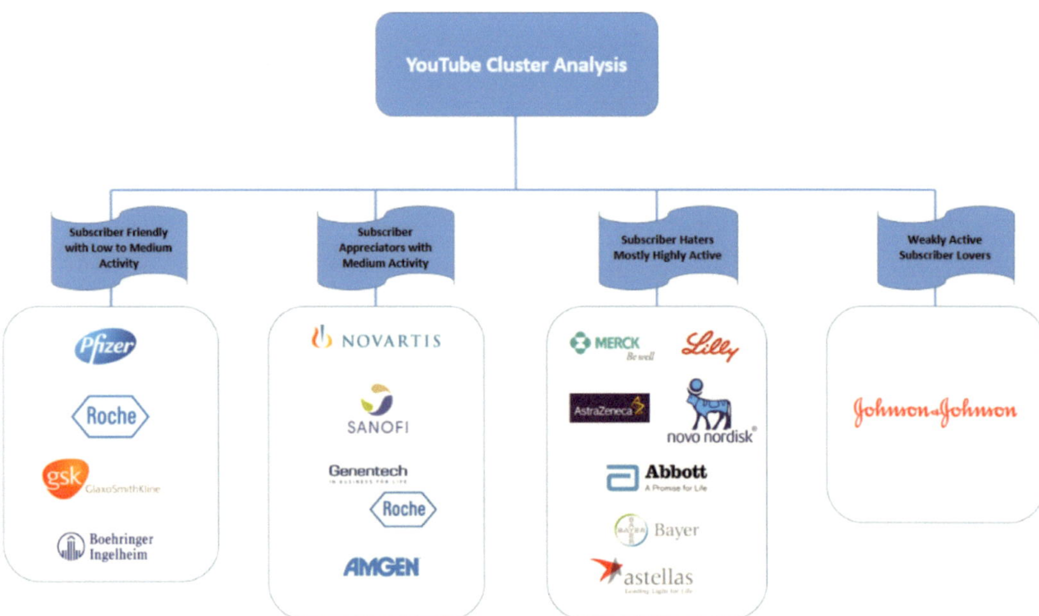

Figure 8. YouTube cluster analysis.

4. Final considerations

Over the last 15 years, social media took over the online world, connecting more people, firms, and brands. However, there are a number of industries that are still lacking in social media exposure and interaction with customers. The question that remains unanswered is what leads to a greater investment in social media by films and in which social media?

In the pharmaceutical industry, social media is a tool that is used to communicate with consumers. However, not all pharmaceutical firms have a Facebook page, Twitter account, or YouTube channel. In fact, only a small number of the pharmaceutical firms analyzed utilize the three social media platforms simultaneously. Also, it seems that smaller pharmaceutical firms are leaving some social media platforms. These observations could be a result of an absence of official guidance in the use of such platforms associated with difficulty to calculate return on investment in the same platforms.

The analysis of the activity of pharmaceutical firms on social media revealed that in the several parameters studied for each social media platform there were diverse sets of scenarios, with some companies performing better than others in one parameter but with a poor performance in others. Also, some companies appeared almost inactive on such platforms. Moreover, when analyzing the ultimate indicator of activity (engagement raking), smaller companies lead the ranking. However, there is not an association between good performance in this ranking with the size of the companies, in terms of revenue and employees. This suggests that other factors could influence digital engagement and therefore, should be evaluated in future research. Regarding digital engagement strategy, pharmaceutical firms do not use the same strategy between the three social media platforms analyzed and have evolved between 2014 and 2017. The cluster analysis showed that in general pharmaceutical firms behave differently on Facebook, Twitter, and YouTube. As their distribution in clusters/segments demonstrated, they present different characteristics among these platforms. However, it seems that Eli Lilly shares some similarities between Facebook and Twitter digital engagement strategy.

The multifaceted communication in the virtual world has recognized consequences besides its connectivity benefits. Pharmaceutical firms are challenged to adopt an increasing digital presence, following major social media trends. Therefore, this work is a small step toward understanding the full potential of social media in this sector, and it leaves an invitation to future research.

Acknowledgements

We gratefully acknowledge the financial support from "Fundação para a Ciência e Tecnologia" (FCT - Portugal), national funding through research grant (UID/SOC/04521/2013) of the Advance/CSG, ISEG and project number UID/ECO/00685/2016 from CEEAplA – from CEEAplA – University of the Azores and from the "Governo dos Açores através da Secretaria Regional do Mar Ciência e Tecnologia/Direção Regional da Ciência e Tecnologia."

Author details

Tiago Costa[1], Teresa Borges-Tiago[2*] and Flávio Tiago[2]

*Address all correspondence to: maria.tp.tiago@uac.pt

1 Unidade de Saúde de Ilha de São Miguel, Ponta Delgada, Portugal

2 University of the Azores, Ponta Delgada, Portugal

References

[1] Applbaum K. Pharmaceutical marketing and the invention of the medical consumer. PLoS Medicine. 2006;**3**:e189

[2] Sharma S, Verma H. Social Media Marketing: Evolution and Change. In: Heggde G, Shainesh G, editors. Social Media Marketing. Singapore: Springer; 2018. pp. 19-36

[3] Levy R. The role and value of pharmaceutical marketing. Archives of Family Medicine. 1994;**3**:327

[4] Fischer M. Marketing Spending Models. Innovation and Marketing in the Pharmaceutical Industry. Springer; 2014. pp. 557-589

[5] Attarabeen O, Alkhateeb FM, Rollins B, Perri M. Pharmaceutical marketing. Jones & Bartlett learning; 2013, 282 pp, $69.95 (softcover), ISBN 9781449697990. American Journal of Pharmaceutical Education. 2013;**77**:135

[6] Tiago MTB, Tiago F, Francisco EBA, Silva S. Healthy 3.0: Healthcare Digital Dimensions. In: Reshaping Medical Practice and Care with Health Information Systems. Ashish Dwivedi; 2016:287-322. [Accessed April 24, 2018]. DOI: 10.4018/978-1-4666-9870-3.ch010

[7] Ding M, Eliashberg J, Stremersch S, editors. The Pharmaceutical Industry: Specificity, Challenges, and What You Can Learn from This Book. In: Innovation and Marketing in the Pharmaceutical Industry: Emerging Practices, Research, and Policies. New York, NY: Springer; 1-18

[8] Buckley J. Pharmaceutical Marketing - Time for Change. EJBO - Electronic Journal of Business Ethics and Organization Studies. 2004;**9**(2). http://ejbo.jyu.fi

[9] Rollins BL. Still the great debate- "fair balance" in direct-to-consumer prescription drug advertising: Comment on "trouble spots in online direct-to-consumer prescription drug promotion: A content analysis of FDA warning letters". International Journal of Health Policy and Management. 2016;**5**:287

[10] Desiraju R, Van Tran T. Spillovers and Other Externalities in Pharmaceutical Marketing. Innovation and Marketing in the Pharmaceutical Industry. Springer; 2014. pp. 673-700

[11] Morton FS, Kyle M. Markets for Pharmaceutical Products. Handbook of Health Economics. Elsevier; 2011. pp. 763-823

[12] Liu Q, Gupta S. Direct-to-Consumer Advertising of Pharmaceuticals: An Integrative Review. Innovation and Marketing in the Pharmaceutical Industry: Springer; 2014. pp. 629-649

[13] Sinkinson M, Starc A. Ask your Doctor? Direct-to-consumer advertising of pharmaceuticals. The Review of Economic Studies; 2015. https://doi.org/10.1093/restud/rdy001

[14] Garattini L, van de Vooren K. Safety and Quality of Generic Drugs: A Never Ending Debate Fostered by Economic Interests? Springer; 2015

[15] Leung D, Law R, van Hoof H, Buhalis D. Social media in tourism and hospitality: A literature review. Journal of Travel & Tourism Marketing 2013;**30**:3-22

[16] Kaplan AM, Haenlein M. Users of the world, unite! The challenges and opportunities of social media. Business Horizons. 2010;**53**:59-68

[17] Aghaei S, Nematbakhsh MA, Farsani HK. Evolution of the world wide web: From WEB 1.0 TO WEB 4.0. International Journal of Web & Semantic Technology. 2012;**3**(1):1

[18] Kotler P, Keller KL. Framework for marketing management. 14th Ed. Pearson Education; 2012. https://www.pearson.com/us/higher-education/product/Kotler-Marketing-Management-14th-Edition/9780132102926.html?tab=features

[19] Kietzmann JH, Hermkens K, McCarthy IP, Silvestre BS. Social media? Get serious! Understanding the functional building blocks of social media. Business Horizons. 2011;**54**:241-251

[20] Tiago T, Amaral F, Tiago F. The good, the bad and the ugly: Food quality in UGC. Procedia-Social and Behavioral Sciences. 2015;**175**:162-169

[21] Zavišić S, Zavišić Ž. Social network marketing. In: Križman-Pavlović, Danijela; Benazić, Dragan, Mijo Mirković, editors. proceedings 22nd CROMAR Congress Marketing challenges in new economy. Hrvatska: University of Pula; 06-08.10.2011. pp. 1008-1019. ISBN: 978-953-7498-45-0

[22] Cheung CMK, Thadani DR. The effectiveness of electronic word-of-mouth communication. A Literature Analysis Electronic Word-of-Mouth Communication. In: proceedings 23rd Bled eConference & eTrust: Implications for the Individual, Enterprises and Society. Bled, Slovenia; 2010:329-345

[23] Tiago T, Veríssimo J. Digital marketing and social media: Why bother? Business Horizons. 2014;**57**:703-708

[24] Lau AY, Gabarron E, Fernandez-Luque L, Armayones M. Social media in health—What are the safety concerns for health consumers? Health Information Management Journal. 2012;**41**:30-35

[25] Hawn C. Take two aspirin and tweet me in the morning: How twitter, Facebook, and other social media are reshaping health care. Health Affairs. 2009;**28**:361-368

[26] Griffiths F, Cave J, Boardman F, Ren J, Pawlikowska T, Ball R, et al. Social networks-the future for health care delivery. Social Science & Medicine. 2012;**75**:2233-2241

[27] Shankar V, Li JK. Leveraging Social Media in the Pharmaceutical Industry. Innovation and Marketing in the Pharmaceutical Industry: Springer; 2014. pp. 477-505

[28] Paul MJ, Dredze M. You are what you tweet: Analyzing twitter for public health. Icwsm. 2011;**20**:265-272

[29] Zheng Y. Patterns and motivations of young adults' health information acquisitions on Facebook. Journal of Consumer Health on the Internet. 2014;**18**:157-175

[30] Greene JA, Choudhry NK, Kilabuk E, Shrank WH. Online social networking by patients with diabetes: A qualitative evaluation of communication with Facebook. Journal of General Internal Medicine. 2011;**26**:287-292

[31] Nabi RL, Prestin A, So J. Facebook friends with (health) benefits? Exploring social network site use and perceptions of social support, stress, and well-being. Cyberpsychology, Behavior, and Social Networking. 2013;**16**:721-727

[32] Hamm MP, Chisholm A, Shulhan J, Milne A, Scott SD, Given LM, et al. Social media use among patients and caregivers: A scoping review. BMJ Open. 2013;**3**:e002819

[33] Stellefson M, Chaney B, Ochipa K, Chaney D, Haider Z, Hanik B, et al. YouTube as a source of chronic obstructive pulmonary disease patient education: A social media content analysis. Chronic Respiratory Disease. 2014;**11**:61-71

Chapter 7

Analysis of GRI Sustainability Reports Issued by Portuguese Public Sector Entities

Maria da Conceição da Costa Tavares

Additional information is available at the end of the chapter

http://dx.doi.org/10.5772/intechopen.76221

Abstract

Issues such as social responsibility and corporate sustainability are now recognized by organizations and the community as very important to achieve sustainable development (SD). Given the increasing pressure from multi-stakeholders, organizations seek to disclose their "best practices" toward SD through a sustainability reporting tool that is prepared on a voluntary basis. Global reporting initiative (GRI) sustainability reports of the Portuguese public sector (PS) entities are used to perform a quantitative longitudinal study with the purpose of identifying the indicators currently disclosed and the GRI application levels. The study focused on the reports of 2008 and 2012. The findings show that Portuguese PS entities report mainly economic indicators, followed by social indicators. Despite the low level of external verification, entities are transparent when declaring their GRI application level.

Keywords: corporate social responsibility, corporate sustainability, sustainability reporting, global reporting initiative, GRI application levels, Portugal, public sector

1. Introduction

As a consequence of recent corporate scandals around the world, companies today face growing pressure from stakeholders to act correctly and to commit themselves to social initiatives (that is, to any program, practice, or policy undertaken by a business firm to benefit society) [1], leading to an increase of concerns about corporate social responsibility (CSR) over the last few decades [2–5], gaining force in international contemporary debates in the last few years [4, 6]. In this sense, different approaches can be found in the academic context to investigate the increasing importance of CSR in society [7]. CSR issues and CSR reports are becoming important, not just nationally, but also globally [3], and CSR has increasingly become more important among business managers, academics, and political decision makers [8].

Companies show their social responsibility by incorporating environmental facts in their management strategies [9]. Facing today's SD challenges, it is accepted that organizations bring about positive change to the world's economic, environmental, and social conditions. As they manage more effectively an issue they can measure, reporting leads to improved SD outcomes [10]. There is a growing tendency among companies to report their sustainability as a way of publicly demonstrating their commitment to the environment and social issues [11]. They seek organizational legitimacy and credibility enhancement by issuing sustainability reports according to the GRI guidelines [12].

Motivated by growing concerns about corporate sustainability and considering the current public pressures for a better behavior the aim of this study is to verify whether Portuguese PS entities reflect good CSR practices in their GRI sustainability reports. Portugal is a European southwestern country, one of the least developed countries in the Eurozone and a small country of the Organization for Economic Co-operation and Development (OECD) [13, 14]. Although in the last years, research has been focusing mainly on the private sector, we can now find a few studies on CSR using PS entities [15]. To the best of our knowledge, this is the first study using the GRI sustainability reports prepared by the Portuguese PS.

Thus, using a longitudinal study, we analyze CSR indicators that are disclosed by Portuguese PS entities in GRI sustainability reports; we also aim to analyze the application levels. The results show that Portuguese PS entities reflect good CSR practices in their GRI sustainability reports as they report mainly economic indicators and despite the low level of external verification, they are transparent when declaring their GRI application level. This paper begins with an approach to CSR and corporate sustainability terms. In the following section sustainability reporting and GRI are explored. Thereafter follow sections of research method, results and discussion. Finally, conclusions, limitations, and areas for further research are presented.

2. Corporate social responsibility and corporate sustainability terms

2.1. Corporate social responsibility

By the end of 1990, the CSR idea became almost universally promoted by all governments, nongovernmental organizations, and individual consumers. CSR has gained emphasis among scholars from a wide variety of subjects and is in vogue, though as a vague concept, with different meanings for different people [16]. CSR is founded on the notion that corporations have relationships with other interests, for instance, with economic, cultural, environmental, and social systems because business activities affect—and are affected by—such interests in society [17].

The most widely used definition of CSR is the one from the Commission of the European Communities in 2001, "a concept whereby companies integrate social and environmental concerns in their business operations and in their interaction with their stakeholders on a voluntary basis," as it integrates five dimensions: voluntary, stakeholders, social, environmental, and economic [16, 18–21]. Truly, a universal definition of CSR is problematic, considering the different national institutional systems of businesses [20] and it is very difficult to define it precisely, as Crane, Matten, and Spence [22] observe when providing an insight into the richness, heterogeneity, and diversity of CSR literature. There are many CSR definitions available [3, 7, 19, 23], although none is widely accepted [20]. As CSR, by definition, is concerned about the responsibilities of companies with regard to other actors in society, it needs to be studied in the context of where it is being practiced [17]. According to CSR studies, corporations use the perception of their activities to influence agents and enhance their image before stakeholders [24].

Some companies consider CSR a negative effect on their business as it may imply costs, in terms of both budget and time. On the other hand, CSR may be seen as positive since it encourages high corporate management, looking closer at the business operations and making them more successful and sustainable in the long term [25]. Dobers and Halme [17], analyzing CSR or SD studies on developing countries or economies in transition, state that little is done. So there is an urgent need for combined efforts from the private sector, PS, and nongovernmental organizations to develop structures and institutions contributing to social justice, environmental protection, and poverty eradication.

It is also apparent that some PS activities, such as procurement, have multiple connections with the contemporary CSR agenda. The PS may choose to address different CSR strategies through actions reflecting a variety of roles: mandating (legislative), facilitating (guidelines on content), partnering (engagement with multi-stakeholder processes), and endorsing (publicity). By using any or a combination of them, a government can seek to increase and improve the level of corporate sustainability reporting [26]. However, the key points in CSR operationalization are its voluntary character and its final aim of enhancing performance in business [27]. Thus, over the decades, the concept of CSR has been growing in importance and significance, being the subject of considerable debate, commentary, theory building, and research. With a broad view of CSR, a firm may enhance its competitive advantage and create win-win relationships with its stakeholders. Additionally, gains from cost and risk reduction and legitimacy and reputation benefits can be achieved [28]. In this context, many experts have noticed the external growth of CSR reporting; few have noticed that its meaning has been internally changing (an exception is Carroll [23]. Future research studies need to redirect the focus to basic research so as to develop conceptual tools and theoretical mechanisms to explain organizational behavior change from a wider social perspective [29].

2.2. Corporate sustainability

As the definition of *sustainability* is pertinent but not widely accepted, Aras and Crowther [30] argue that the definition in the Brundtland Report [31] must be seen as a starting point since there is a clear agreement: it was with this report, under the title "Our Common Future," that the sustainability concept and essence were popularized [32]. This is the original SD or sustainability concept [32–36]. It is obvious that the terms *sustainability* and *SD* are used as equivalent and seen by many as synonyms [37].

The SD concept combines economic prosperity, a better environment, and social justice aims, which demand an integrated strategy allowing for practical measures to achieve a better quality of life for people now and in the future [29, 38]. It supports a balance between present and future needs, although it does not specify them or define the balance to be implemented [39]. The United Nations has stated that SD can only become a reality if corporate responsibility becomes a dominant concern for individual companies and the business community as a whole [40].

Although SD is a societal concept, it is increasingly being applied as a corporate concept under the name of *corporate sustainability* [41]. The word *sustainability* is one of the most widely used words related to corporate activity [37], despite being a controversial term, as it means different things for different people [30, 39]. Sustainability requires a collective decision-making level for the common good [42], and any definition of *sustainability* should cover what is known as intergenerational equity [39]. Sustainability is focused on the future, which necessarily implies the acceptance of all the costs involved at the present as an investment for the time to come [30]. This is mainly a global concept emphasizing not only an efficient allocation of resources throughout time but also a fair distribution of resources and opportunities among current, present, and future generations [42, 43]. Sustainability is often articulated in terms of the tripartite model (economic, environmental, societal). Regarding a community, sustainability is considered in terms of four fundamental and closely related themes: ethics, conservation, cooperation, and competition [44].

Marrewijk ([45], p. 95) shows that definitions of *CSR* and *corporate sustainability*—"one solution fits all"—should be abandoned, "accepting various and more specific definitions matching the development, awareness and ambition levels of organizations." CSR as a new tool fits into the current corporate responsibility or corporate sustainability framework to complete the image of corporate sustainability. In general, corporate sustainability and CSR refer to company activities—voluntary by definition—demonstrating the inclusion of social and environmental concerns in business operations and interactions with stakeholders. Aras and Crowther [4] argue that four aspects of sustainability must be considered as the key dimensions of sustainability that need to be recognized and analyzed: societal influence, environmental impact, organizational culture, and finance. These four aspects can be resolved into a two-dimensional matrix along the polarities of internal versus external focus and short-term versus long-term focus, which together represent a complete representation of organizational performance. The company is firmly embedded into a global environment that necessarily takes into account the past and the future as well as the present. A short-term approach is no longer acceptable for sustainability as it pays attention to the future as well as to the present [30].

Organizations adopting sustainability as part of their corporate culture explore triple bottom line (TBL) as part of their business strategy and simultaneously create value for all their stakeholders [46]. Corporate sustainability, as a building ideology for rethinking business, requires systemic corporate cultural changes, engaging all stakeholders and building a sustainable society as part of it. And the fundamental premise of corporate sustainability is that organizations should fully combine social and environmental objectives with financial ones and explain their well-being actions to a wider range of stakeholders through an accountability and reporting mechanism [47].

3. Sustainability reporting and GRI

3.1. Sustainability reporting: background

Historically, sustainability reporting, in the strictest sense of the word, was preceded by three different types of reporting: annual, environmental, and social. Then emerged "sustainability reporting" as a designation for this new integrated form of economic, environmental, and social reporting [48]. According to Sciulli [49], p. 76, a new phase of research opportunities' expansion has come up, and the last tendency seems to favor sustainability reporting, a term that seems to have replaced the "phrase social and environmental accounting research" and implies an emphasis on organizations seeking to report more information than it is included in traditional financial accounting. In this reporting, there are broader techniques of sustainability accounting and accountability that have the potential to be powerful tools in the management, control, and accountability of organizations for their social and environmental impacts [40]. Thus, social, ethical, and environmental reporting is aimed at different stakeholders and is assumed to spread a company's accountability beyond financial accounting, understanding that organizations do not solely have financial responsibilities but also social, ethical, and environmental ones, which should be used to ascertain organizations' accountability [50, 51]. However, CSR reports are not new, and a lot of companies have been preparing them under several inherent titles. Initially, those reports may have a public relations appearance for companies, with a positive interpretation of their results. However, with their evolution together with the issues raised by several stakeholders, these reports have come up with more quantifiable targets and results presentation [46]. CSR reporting is, then, an important aspect of social and environmental accountability [52].

Regarding the terminology for reporting and according to KPMG [53], it varies globally between companies: "sustainability" reporting (43%), "corporate social responsibility," (25%) and "corporate responsibility" (14%). In Zorio et al.'s [11] and Skoulodis and Evangelinos' study [54], CSR reporting and sustainability reporting are used as synonyms, referring to reports presenting economic, environmental, and social aspects of corporate activities and emerging as a new corporate reporting tendency. These reports describe policies, plans, and programs the company puts into practice, including quantitative and qualitative information on economic, environmental, and social performance, which Elkington [55] has described as the company's TBL in a stand-alone publication [54, 56, 57].

According to Owen [58], there have been several attempts to establish a global common framework for CSR reporting, which covers mostly economic, social, environmental, and governance dimensions [3]. Actually, corporate reporting, which used to be designated as environmental reporting, and later as CSR reporting, is now repackaged as sustainability reporting [4].

In this sense, several definitions of corporate sustainability reporting are available in published literature, though there is none that is universally accepted [41]. Milne and Gray [57], by tracing the history of the evolution of corporate sustainability reporting, identify and isolate the TBL concept as a core and dominant idea. Additionally, this process has become reinforced and institutionalized through KPMG's triennial surveys of practice.

Sustainability reporting is the action through which an organization publicly communicates its economic, environmental, and social development as a routine and comparable to organizations' financial reports [10]. It is a way of helping organizations inform on their performance and enhance their accountability [35], integrating this information in a single publication, which is gaining acceptance among a growing number of organizations [54]. Since sustainability reporting is a somewhat new practice—disclosures are expected to increase over time—because of lack of research focusing on sustainability issues in the PS [49].

According to Haque et al. [59], the PS as an organizational system has components similar to private organizations: leadership, strategic planning, communication and coordination, administrative procedures, and public responsibility. There are several authors approaching these SD issues in the PS, namely, Burritt and Welch [60]; Larrinaga-González and Bebbington [61]; Ball [62, 63]; Ball and Grubnic [64]; Ball and Bebbington [65]; Broadbent and Guthrie [66]; Guthrie and Farneti [40]; Larrinaga-González and Pérez-Chamorro [67]; Lewis [68]; Burritt and Schaltegger [69]; Sciulli [70]; and Gray and Laughlin [71]. However, despite the new legislative guidelines for "Good Governance Practices" [72] sustainability reporting according to the GRI guidelines, of a voluntary nature, is recent in the Portuguese PS.

3.2. Global reporting initiative

The GRI was created at the end of 1997 from a project managed and financed by the Coalition for Environmentally Responsible Economies (CERES) [32, 73, 74]. Since the introduction of the CERES Principles in 1989, sustainability reports have been the main tool companies use to show the outside world their social responsibility [73]. The GRI's mission is to offer a reliable structure for sustainability reporting, with a globally shared structure of concepts, a consistent language, and a largely understood metric to communicate issues related to sustainability in a clear and transparent way, which may be used by several organizations regardless of their dimension, sector, or location [75, 76]. This is to elevate sustainability reporting to a similar level as financial reporting in terms of comparability, rigor, auditability, and general acceptance [77].

The GRI's explicit objective is to enlighten and harmonize nonfinancial reporting [10, 73], and its main activity is to develop and promote a coherent framework for this reporting [78]. The GRI has tried to broaden its (global) range, scope (social, economic, and environmental performance indicators), flexibility (descriptive and quantitative indicators), and stakeholder base (industry, financial sector, accounting, civilian, environmental society and nongovernmental organizations of human rights, work, among others) [73]. The GRI claims to supply the entire world with a standard base of comparable reports on sustainability, that is, generic SD indicators between the three sustainability dimensions (or TBL) [79, 80], a concept introduced by John Elkington in 1994 [73, 76, 81, 82]. Since its conception in 1999, the GRI has become a model leader in voluntary sustainability reporting, producing a guidelines framework for sustainability reporting. This is a prominent framework for voluntary corporate reporting on environmental and social performance all over the world, and it is generally considered very successful [73, 83, 84]. And while sustainability reporting is a voluntary process, companies will not discharge accountability [85]. However, Lynch [86] argues that the low level of reporting under the GRI guidelines is disappointing. The fourth generation of the GRI guidelines (G4) proposes alterations on the information on management, new orientations for defining the report limits, and new information to be reported in key areas, such as governance and supply chains. Its mission is that these reports publication becomes a standard practice, offering orientation and support to organizations, allowing a greater comparability between reports and companies within the same sector [87].

In the PS, the GRI is the predominant framework [88–90], providing a vision for SD [91]. The GRI argues that the PS has a great impact on the national and global progress toward SD [92–94]. The GRI argues that the PS has the civic responsibility of properly managing public assets, resources, and/or facilities in such a way that it supports SD aims and a public and transparent report of its activities to promote sustainability [40, 93]. An effective performance in the PS is frequently driven more by strong organizational cultures, good management practices, and effective communication networks rather than by rules and regulations or procedures and salary tables [95]. "Sustainability reporting is a key tool for demonstrating the role of public agencies in advancing sustainable development" ([96], p. 328).

3.3. GRI application level criteria

Few studies have analyzed the factors influencing the application level of GRI indicators [16, 40, 90, 97], as well as quality, transparency and credibility in sustainability disclosure [98–104]. The GRI application levels were introduced in 2006, with the launching of the G3 Guidelines. Therefore, as far as the GRI is concerned, the quality of information reported must be established on comparability, reliability, clarity, balance, accuracy, and timeliness principles. The application levels show the extent to which the GRI's framework has been applied in a sustainability report, and they communicate which disclosure items from the guidelines or sector supplements have been addressed. In a report based on the GRI guidelines, organizations should report the level to which they have applied the GRI reports framework through the "application levels" systems (see self-declaration of GRI application levels (2000–2011) [105]).

To respond to beginner, intermediate, and advanced reporters, the system presents three levels, titled C, B, and A. The reporting criteria in each level indicate the evolution. The levels are related to the number of items and the set of addressed GRI "report content." An organization may self-declare an extra point (+) in each level (for example, C+, B+, or A+) if the report was audited by an external entity and/or GRI. A key point to note is that a report's application level is self-declared by the reporting organization. Organizations can choose to sign up for the GRI Application Level Check to confirm their understanding of the application level system [32, 87, 106]. The formalization of these different levels of application of the GRI framework is supposed to facilitate the reliability assessment of the reports and to strengthen their transparency, so that, in theory, higher application levels of the GRI reports (A + e A) are supposed to mitigate the uncertainty and the credibility gap associated with mistrust toward information on sustainable development reported by organizations [98, 99].

4. The research method

The central questions used to guide this study were: which TBL indicators are disclosed by Portuguese PS entities in GRI sustainability reports? Do GRI sustainability reports equally reflect the TBL dimensions? And which are the GRI application levels?

To address these questions, a longitudinal study was used for Portuguese PS entities that issued sustainability reports according to the GRI guidelines in 2008 and 2012. Case studies are particularly suitable for exploratory case studies focused on the study of emergent practices [107]. Case studies of longitudinal nature can elicit a great deal of data over a period of time [108]. "An interpretive and preferably longitudinal case study approach would thus seem to be a useful research strategy, adding 'flesh' to the theoretical 'skeleton'" ([109], p. 301).

KPMG International argues that the use of the GRI guidelines is almost universal: 78% of reporting companies worldwide use GRI reporting guidelines in their corporate responsibility reports, a rise of 9 points since the 2011 survey (over 90% in South Korea, South Africa, *Portugal*, Chile, Brazil, and Sweden) [53]. In Portugal, the rate of corporate responsibility reporting was of 52% in 2008, 69% in 2011, and 71% in 2013, according to KPMG's survey [53, 110].

In this study, 58 GRI sustainability reports disclosed by PS organizations were collected based on a review of the GRI database and/or on the BCSD Portugal website and/or on the entities' website and/or using the search engine "google.pt." There has been content analysis to observe and identify the information elements of the economic, social, and environmental performance and GRI application levels. With the aim of understanding the TBL indicators that are disclosed in GRI sustainability reports and the application levels of Portuguese PS GRI sustainability reports, the data were analyzed using the Statistical Package for the Social Sciences (SPSS) version 23.0.

4.1. Background information on the entities in the sample

The study focused on Portuguese PS sustainability reports following the GRI guidelines, with data from 2008 and 2012. In the year 2008 there was a significant increase of publications, including in the PS, and there were, for the first time, publications from the administrative PS. The year 2012 was chosen as it was when, after a decrease, the number of publications rose again, and for the second time, there were publications with information from the administrative PS.

The sample is composed of 58 reports of PS entities, and of these, only two in 2008 and five in 2012 have a different title from "sustainability report," although the term *sustainability* is used. **Figure 1** presents background information on the entities included in the sample. PS entities are classified as government business enterprises (GBEs) and administrative PS entities, "aggregated" into nine industries. This classification was based on the activities developed by each entity.

As it can be noticed, the "transportation" and "water and waste management" industries represent more than 50% of the sample (59.4% in 2008 and 73.1% in 2012), which represents 65.6% (38 reports) of the sample (19 in 2008 and 2012). The number of reports reduced in 2012 in most of the industries (from 32 to 26). In an economic crisis context, the PS has focused on reducing costs and increasing revenues, concerned about economic stability and sustainability, leading to a decrease of their sustainability reporting strategies. The administrative PS presents the fewest industries, represented in 2008 by "local government," with four entities, and in 2012, by "local government," with one entity, and by "education," with one entity (10.3% of the sample).

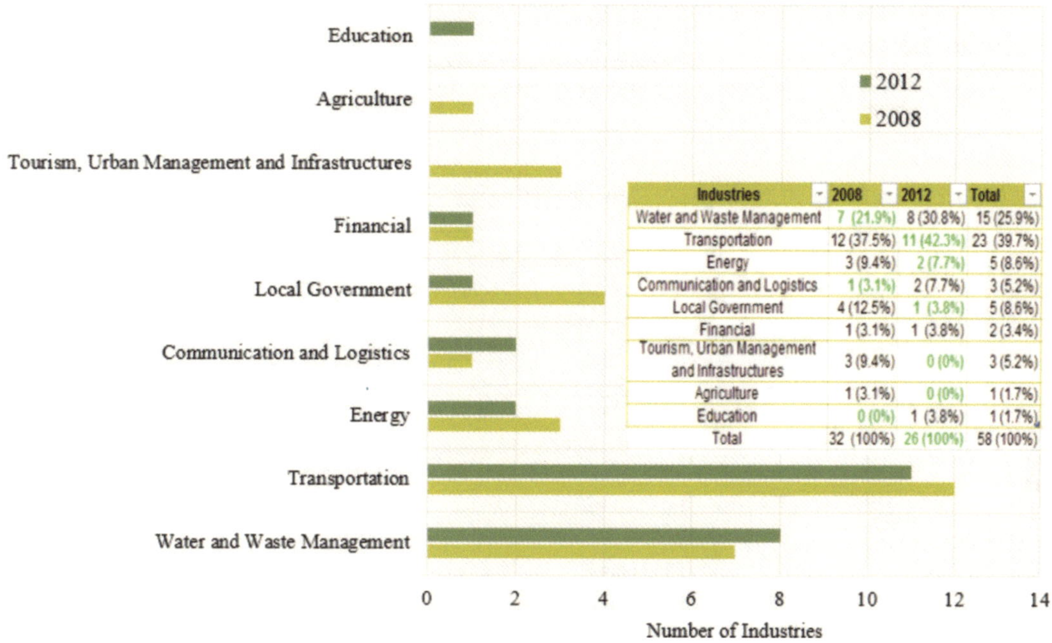

Industries	2008	2012	Total
Water and Waste Management	7 (21.9%)	8 (30.8%)	15 (25.9%)
Transportation	12 (37.5%)	11 (42.3%)	23 (39.7%)
Energy	3 (9.4%)	2 (7.7%)	5 (8.6%)
Communication and Logistics	1 (3.1%)	2 (7.7%)	3 (5.2%)
Local Government	4 (12.5%)	1 (3.8%)	5 (8.6%)
Financial	1 (3.1%)	1 (3.8%)	2 (3.4%)
Tourism, Urban Management and Infrastructures	3 (9.4%)	0 (0%)	3 (5.2%)
Agriculture	1 (3.1%)	0 (0%)	1 (1.7%)
Education	0 (0%)	1 (3.8%)	1 (1.7%)
Total	32 (100%)	26 (100%)	58 (100%)

Figure 1. Number of entities per industry and year.

5. Results

The results are presented below in subsections. One section briefly presents details of the indicators presented in the reports, and another examines GRI application levels.

5.1. The TBL indicators in the reports

The sustainability indicators set by the GRI (G3/G4) guidelines are divided into three categories: economic (7), environmental (17), and social (25), with a total of 49 essential indicators (100%). **Figure 2** presents the descriptive statistics of TBL dimensions in the 2 years under study.

The sample reveals that economic indicators have ranged between 43% and 100%, with a mean of 80.2% in 2008 and 87.3% in 2012. Environmental indicators ranged between 50 and 100%, with an average of 73.1% in 2008 and 80.8% in 2012. Social indicators ranged between 31 and 100%, with an average of 75.2% in 2008 and 83.6% in 2012. Both in 2008 and in 2012, the economic indicators came up in the first place, followed by social indicators. However, in 2012, the values of the three indicators were greater than the ones in 2008.

Since 2007, Portugal has been one of the European Union members most affected by the global financial crisis [111], and this can be the explanation for this result. The financial crisis may lead organizations to move away from the socially responsible behavior as it costs a lot to

Aggregation of Industries		% Economic Indicators		% Environmental Indicators		% Social Indicators		Reports
		Mean	s	Mean	s	Mean	s	N
Water and Waste Management	2008	95,9	11,0	87,4	18,7	93,1	14,7	7
	2012	96,4	10,3	90,4	17,2	93,0	18,2	8
Transportation	2008	70,3	29,6	67,2	24,0	67,0	30,9	12
	2012	75,3	28,0	68,5	28,1	71,6	33,2	11
Energy	2008	95,3	8,1	98,0	3,5	98,0	2,3	3
	2012	100,0	0,0	100,0	0,0	100,0	0,0	2
Communication and Logistics	2008	43,0		65,0		80,0		1
	2012	85,5	20,5	94,0	8,5	93,0	9,9	2
Local Government	2008	82,0	13,9	50,0	3,5	31,0	14,0	4
	2012	100,0		53,0		72,0		1
Financial	2008	100,0		100,0		100,0		1
	2012	100,0		100,0		100,0		1
Tourism, Urban Management and Infrastructures	2008	81,0	21,9	64,7	23,5	96,0	4,0	3
Agriculture	2008	71,0		94,0		88,0		1
Education	2012	100,0		82,0		84,0		1
Total	2008	80,2	23,8	73,1	23,2	75,2	29,7	32
	2012	87,3	22,0	80,8	24,2	83,6	26,0	26

Legend: < 50%; [50%–75%]; > 75%

Figure 2. Descriptive statistics per category, industry, and year.

meet stakeholder's expectations [112]. The variation observed in 2012 may be one explanatory and differentiating factor in the inclusion of environmental and social concerns in organizations. Although an economic and financial crisis, social responsibility makes them less vulnerable because it is a tool associated with the fulfillment of legal obligations and organizations' "good practices." These are too important in maintaining their reputation and competitive advantage, even during a period of financial crisis, as Rodrigues et al. [111] state. This period is an opportunity to restore or improve the image and levels of business confidence, because "society and the community are perceived to be stakeholders whose needs deserve greater urgency and stronger legitimacy explanations" ([113], p. 667). Organizations increase their CSR performance to build or sustain their brand name, consumers' trust and redefine the relationship between companies and society. Thus, the crisis gives companies the opportunity to redirect CSR, which is transforming a threat into an opportunity [112].

5.2. The GRI application levels

With the objective of analyzing the GRI (G3) application levels, undeclared, self-declared (C, B, A), external verification (C+, B+, A+), verified by GRI, all 58 reports were encoded using an 8-point scale, where 0 = undeclared application level, 1 = application level C, 2 = application level B, 3 = application level A, 4 = application level C+, 5 = application level B+, 6 = application level A+, and 7 = verification GRI. **Table 1** highlights how the application levels of GRI sustainability reports of the sample are distributed per industry.

http://dx.doi.org/10.5772/intechopen.76221

Industry aggregation	Year	Undeclared	Application Level							Total
			Self-declared			External verification				
			C	B	A	C+	B+	A+	GRI	
WWM	2008	0	0	0			3	4	0	7
	2012	1	0	2	3			1	1	8
	Total	1	0	2	3		3	5	1	15
T	2008	5	3	4			0	0	0	12
	2012	4	2	2	2			1	0	11
	Total	9	5	6	2		0	1	0	23
En	2008	0	0	0			1	0	2	3
	2012	0	0	0	0			2	0	2
	Total	0	0	0	0		1	2	2	5
CL	2008	0	0	0			1	0	0	1
	2012	0	0	1	0			1	0	2
	Total	0	0	1	0		1	1	0	3
LG	2008	4	0	0			0	0	0	4
	2012	1	0	0	0			0	0	1
	Total	5	0	0	0		0	0	0	5
F	2008	0	0	0			0	1	0	1
	2012	0	0	0	0			1	0	1
	Total	0	0	0	0		0	2	0	2
TUI	2008	1	2	0			0	0	0	3
	2012									
	Total	1	2	0	0		0	0	0	3
Ag	2008	1	0	0			0	0	0	1
	2012									
	Total	1	0	0	0		0	0	0	1
Ed	2008									
	2012	1	0	0	0			0	0	1
	Total	1	0	0	0		0	0	0	1
Total	2008	11	5	4			5	5	2	32
	2012	7	2	5	5			6	1	26
	Total	18	7	9	5		5	11	3	58

Legend: WWM = water and waste management; T = transportation; En = energy; CL = communication and logistics; LG = local government; F= financial; TUI = tourism, urban management, and infrastructures; Ag = agriculture; Ed = education.

Table 1. GRI application level per industry.

In the 2 years studied, out of the 23 reports from the "transportation" industry, 9 chose not to declare their level, 13 self-declared it, and 1 did it by external verification (A+). Out of the 15 reports, "water and waste management," in 2008, all opted for the external verification (3 B+ and 4 A+). In 2012, out of the eight entities, one chose not to declare its level, five self-declared it, and two declared external verification (1 A+ and 1 GRI). In the "energy" industry, all the entities (five in both years) opted for external verification (1 B+, 2 A+, and 2 GRI). In the "local government" industry, all the entities disclosing in both years (five reports) chose not to declare their level. In communication and logistics," in 2008, one entity chose the external verification, and in 2012, one chose the self-declaration, and one external verification (A+). In "tourism, urban management, and infrastructures," in 2008, one entity did not declare and two self-declared. The "financial" entity opted for external verification in both years (A+). "Agriculture," in 2008, and "education," in 2012, did not declare the application level. "Water and waste management" and "energy" were the entities where most chose the external verification of the disclosure level of their reports and those that are assessed by the GRI. A total of 36% of the entities studied opted for self-declaring their application level, 33% opted for external verification, and 31% for not declaring it. **Figure 3** presents the industries' GRI application level of the entities under study.

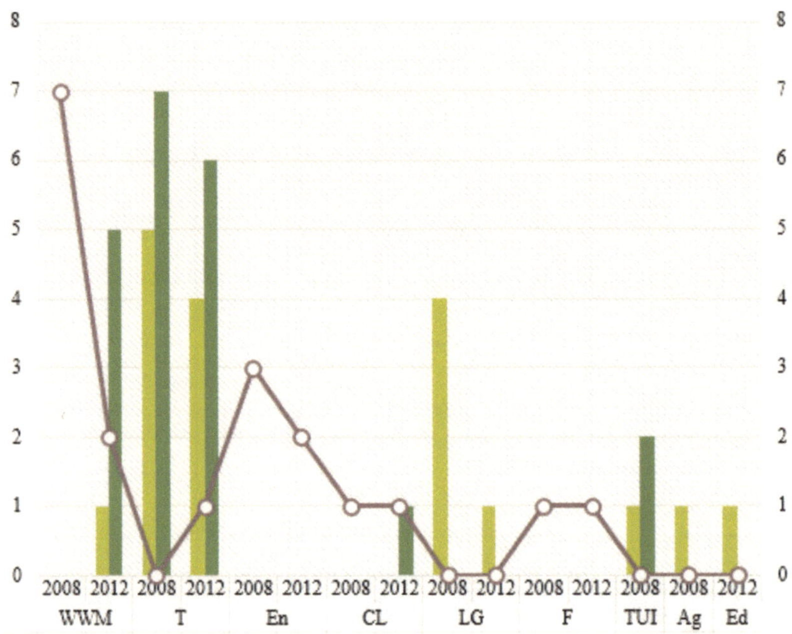

	WWM 2008	WWM 2012	T 2008	T 2012	En 2008	En 2012	CL 2008	CL 2012	LG 2008	LG 2012	F 2008	F 2012	TUI 2008	Ag 2008	Ed 2012
undeclared	0	1	5	4	0	0	0	0	4	1	0	0	1	1	1
self-declared	0	5	7	6	0	0	0	1	0	0	0	0	2	0	0
external verification	7	2	0	1	3	2	1	1	0	0	1	1	0	0	0

Legend: WWM = water and waste management; T = transportation; En = energy; CL = communication and logistics; LG = local government; F = financial; TUI = tourism, urban management, and infrastructures; Ag = agriculture; Ed = education.

Figure 3. GRI application level.

Considering the total of industries in the 2 years, the application levels of external verification were of 9/15 (60%) in "water and waste management," decreasing in 2012; of 1/23 (4%) in "transportation," increasing in 2012; of 5/5 (100%) in "energy"; of 2/3 (67%) in "communication and logistics"; of 0/5 (0%) in "local government"; of 2/2 (100%) in "financial"; of 0/3 (0%) in "tourism, urban management, and infrastructures"; of 0/1 (0%) in "agriculture"; and of 0/1 (0%) in "education." Summing up, there were 12/32 (37.5%) external verifications in 2008 and 7/26 (26.9%) in 2012.

Summing up, first, the results of the 58 sustainability reports studied, organized into nine industries, show that the three TBL dimensions, according to the GRI guidelines, are widely disclosed, although the indicators vary between industries. They mostly present values above 75%, despite some supremacy of economic indicators, followed by the social ones and at last by the environmental ones. They report on the three TBL areas, although the extension of disclosure varies according to the industry where the entity operates, as found by Roca and Searcy's study [41]. All areas of the TBL were widely disclosed by Portuguese PS entities in their GRI sustainability reports, and this disclosure increased from 2008 to 2012.

Second, in terms of the application levels, there are a significant number of entities that opted for not declaring, and most of them opted for self-declaring their application level. This fact may be related to the analyzed period of a severe financial crisis. However, the external verification would have legitimized their action and the risk of reputation of their activities. Although this authentication is not mandatory by a third party, this procedure represents the answer to the demands from stakeholders and reinforces the credibility, reliability, and transparency of both organizations and the GRI [98–101].

6. Discussion

This article explores which TBL indicators are disclosed by Portuguese PS entities in GRI sustainability reports and which are GRI application levels.

As noted in the literature review, Ball [63] found that accounting - social and environmental - is pressed into use to promote a change toward SD. However, researchers still struggle with the definition of SD and with its key determinants [100, 114]. Roca and Searcy [41] observe that names such as "sustainability," "sustainable development," "corporate social responsibility," "corporate responsibility," "triple bottom line" and "accountability" reports, among many others, are used to refer to sustainability reports. Also, according to KPMG [53], the term *corporate responsibility* includes the concept of "sustainability."

In this sense, the empirical results of this study show that all three areas of the TBL indicators are, in general, widely addressed in GRI sustainability reports in Portuguese PS entities, which supports the definitions of CRS, corporate sustainability, and sustainability reporting mentioned earlier and highlighted by literature.

The study of Giannarakis and Theotokas already indicates organizations have increased CSR performance before and during the financial crisis (except for the period 2009–2010), in order to regain the lost trust in businesses. The investment view of CSR can help organizations differentiating their goods or services and re-establishing the trust between organizations and their stakeholders. The benefits that may arise by the implementation of CSR strategy and initiatives are more important than ever for the organizations' survival [112].

On the one hand, these findings give credibility to the argument that GRI is becoming an established institution and provides structure and guidance to the report as supported by Boiral and Henri [100], Godha and Jain [101], Denčić-Mihajlov and Zeranski [102], Brown et al.'s [115] and Antoni and Hurt's [84], for example. The use of the GRI framework, that proposes detailed guidelines on how to consider the economic, social, and environmental dimensions of SD, allows organizations not only to understand the concept of SD better, which is rarely clearly defined, but also the manner of its implementation [100].

On the other hand, the question why the number of entities reporting under these guidelines is still so low in Portugal comes up. It is believed that the differences in organizations' resources availability may contribute to the lack of social responsibility disclosure suggested by GRI guidelines.

This paper corroborates Antoni and Hurt [84], who emphasize that sustainability reporting is a shortfall, and Guthrie and Farneti [40], Lewis [68] and Sciulli [49], who assert that this practice is still in infancy in the PS. In addition, there is also an agreement with Ball and Grubnic [64], when they state that the PS presents a transformative potential of sustainability accounting and accountability.

In fact, CSR public policies adopted by governments to promote responsible and sustainable business practices neither gives an answer to the needs of today's societies nor makes it possible to understand the new challenges facing social governance in depth, as Albareda et al. [18] state. Thus, González and Martinez [6] verify that the existence of a regulatory framework and other policies to promote CSR would also be important. It also seems crucial the role of a key individual within each organization that would lead the PS to report, as Farneti and Guthrie [88] affirm. In effect, disclosures can be related with organizational strategies and operational activities, consistent with the findings of previous studies of Larrinaga-González and Pérez-Chamorro [61] and Lewis [68].

Just as Lynch [86], it is considered that there is capacity for improving reporting practices and that the government's leadership and action could be an important driver to the adoption of sustainability reporting. Also, mandatory GRI adoption would allow comparison over time. Moreover, Sciulli [70] is also corroborated on the opinion that local government leadership together with communication with stakeholders and community engagement are able to influence sustainability reporting.

7. Conclusion

This study has contributed toward addressing a research gap in PS sustainability reporting by providing an initial understanding of current sustainability reporting practices in the PS in Portugal. It was found that Portuguese PS entities do not face a number of pressures to produce sustainability reports nor to have their reports evaluated by an independent and skilled third party, to legitimize their activities. Still, sustainability issues are not yet actively considered within the entities' strategic plans and practices. However, it is considered that the disclosing entities tend to be recognized for good reporting practices, as those which were early adopters, which have a better understanding of these issues, and experience and learning.

In fact, there are relatively few published examples of the actual use of sustainability indicators and GRI application levels in Portuguese PS entities. Answering this study's questions, an insight into TBL indicators in GRI sustainability reports was provided as well as the way these tools are used by the PS for a greater transparency of its activities.

The research showed that the indicators disclosed were relatively well distributed along the three dimensions of TBL of sustainability, despite some supremacy of economic indicators. However, the entities under study have a low level of external verification. A significant number of entities self-declared a certain level, based on their own assessment of the report content, when compared with the criteria of the GRI application levels. Other entities have asked for an external entity of assessment to give an opinion about the self-declaration and/or asked the GRI to examine their self-declaration. This certification acknowledges that the information disclosed is true and accurate. Given the continued growth in the application of the GRI guidelines worldwide, the research also yielded further insight into the actual disclosure of the GRI indicators.

In this sense, this work tries to answer Cerin and Scholtens's [114] and Lee's [29] calls for future investigations in CSR. Cerin and Scholtens [114] are also supported when they point out the lack of a coherent theoretical framework for SD. Thus, SD and CSR research should continue to be studied from a wide variety of theories and perspectives. "Maybe one day we shall witness a paradigm switch and a new discipline (sustainomics, sustainology, sustainosophy?) may arise" ([114], p. 72). Ball and Bebbington's message [65] is upheld when stating that the PS's distinctive profile and particular opportunities can support society's pursuit on accounting and reporting for SD. Thus, traditional accounting, although still pivotal, is not sufficient and organizations have to consider disclosing information that addresses other aspects, such as social and environmental issues.

The research is of interest to academicians and practitioners who are interested in the theory and practice of sustainability reporting or TBL reporting [32]. And there are numerous possibilities for future research in this area, especially in the PS.

It is important to understand why the disclosure of social responsibility and corporate sustainability "good practices" is still so incipient. Despite legal orientations regarding the duty of disclosing those accounting practices and the existence of guidelines from international entities such as the GRI, voluntary social responsibility and sustainability disclosure practices, according to the GRI tool to sustainability reporting, are still reduced. Thus, this is a fascinating and worthy-of-study issue.

Case studies could provide insight into the process of developing, implementing, using, and improving indicators over time. The disclosure of other parameters of indicators could be explored. Questionnaires could be used to explore in greater depth how the usefulness of the GRI indicators is perceived by entities. Research on the determinants of the indicators' disclosure in different sectors may help further explain how indicators are selected and used. Interviews would allow corporate managers to explain their approach on many questions, such as lack of external verification and factors influencing this decision taking, leading entities to ask for an audit of their sustainability reports, validating the importance of this process for the credibility and reputation of the reporting entities. In future research, the use of indicators in the public and private sectors could be compared. Finally, research on mandatory and voluntary reporting can also be a line for future work.

Author details

Maria da Conceição da Costa Tavares

Address all correspondence to: mariatavares@ua.pt

University of Aveiro, Aveiro, Portugal

References

[1] Brønn PS, Vidaver-Cohen D. Corporate motives for social initiative: Legitimacy, sustainability, or the bottom line? Journal of Business Ethics. 2009;**87**:91-109

[2] Campbell JL. Why would corporations behave in socially responsible ways? An institutional theory of corporate social responsibility. Academy of Management Review. 2007;**32**(3):946-967

[3] Golob U, Bartlett JL. Communicating about corporate social responsibility: A comparative study of CSR reporting in Australia and Slovenia. Public Relations Review. 2007;**33**:1-9

[4] Aras G, Crowther D. Corporate sustainability reporting: A study in disingenuity? Journal of Business Ethics. 2009;**87**:279-288

[5] Carnevale C, Mazzuca M. Sustainability reporting and varieties of capitalism concetta. Sustainable Development. 2012;**2**:361-376

[6] González MC, Martinez CV. Fostering corporate social responsibility through public initiative: From the EU to the Spanish case. Journal of Business Ethics. 2004;**55**(3):275-293

[7] Schultz F, Wehmeier S. Institutionalization of corporate social responsibility within corporate communications: Combining institutional, sensemaking and communication perspectives. Corporate Communications: An International Journal. 2010;**15**(1):9-29

[8] Skouloudis A, Evangelinos K, Kourmousis F. Assessing non-financial reports according to the global reporting initiative guidelines: Evidence from Greece. Journal of Cleaner Production. 2010;**18**:426-438

[9] Sarmento M, Durão D, Duarte M. Study of environmental sustainability: The case of Portuguese polluting industries. Energy. 2005;**30**:1247-1257

[10] Leyira CM, Uwaoma IE, Olagunju A. Corporate social responsibility and compliance with regulations in Nigeria. Research Journal of Finance and Accounting. 2012;**3**:35-41

[11] Zorio A, García-Benau MA, Sierra L. Sustainability development and the quality of assurance reports: Empirical evidence. Business Strategy and the Environment. 2013;**22**:484-500

[12] Hedberg C, Malmborg F. The global reporting initiative and corporate sustainability reporting in Swedish companies. Corporate Social Responsibility and Environmental Management. 2003;**10**:153-164

[13] Lopes PT, Rodrigues LL. Accounting for financial instruments: An analysis of the determinants of disclosure in the Portuguese stock exchange. The International Journal of Accounting. 2007;**42**(1):25-56

[14] Branco ML, Rodrigues LL. Factors influencing social responsibility disclosure by Portuguese firms. Journal of Business Ethics. 2008;**83**:685-701

[15] Mathews M. Twenty-five years of social and environmental accounting research: Is there a silver jubilee to celebrate? Accounting, Auditing and Accountability Journal. 1997; **10**(4):481-531

[16] Gallego I. The use of economic, social and environmental indicators as a measure of sustainable development in Spain. Corporate Social Responsibility and Environmental Management. 2006;**13**:78-97

[17] Dobers P, Halme M. Corporate social responsibility and developing countries. Corporate Social Responsibility and Environmental Management. 2009;**16**:237-249

[18] Albareda L, Lozano JM, Ysa¸ T. Public policies on corporate social responsibility: The role of governments in Europe. Journal of Business Ethics. 2007;**74**(12):391-407

[19] Dahlsrud A. How corporate social responsibility is defined: An analysis of 37 definitions. Corporate Social Responsibility and Environmental Management. 2008;**15**:1-13

[20] Jackson G, Apostolakou A. Corporate social responsibility in Western Europe: An institutional mirror or substitute? Journal of Business Ethics. 2010;**94**:371-394

[21] European Comission. [e-book] LIVRO VERDE – Um quadro para as políticas de clima e de energia em 2030. 2013. Available from: http://www.europarl.europa.eu/meetdocs/2009_2014/documents/com/com_com(2013)0169_/com_com(2013)0169_pt.pdf [Accessed: December 12, 2015]

[22] Crane, A, Matten, D, Spence, LJ. Corporate social responsibility: In a global context. In: Crane A, Matten D, Spence LJ, editors. Corporate Social Responsibility: Readings and Cases in a Global Context. 2nd ed. Abingdon: Routledge; 2013, pp. 3-26. [Online]. Available from https://papers.ssrn.com/sol3/papers.cfm?abstract_id=2322817 [Accessed: October 22, 2016]

[23] Carrol AB. Corporate social responsibility: Evolution of a definitional construct. Business & Society. 1999;**38**(3):268-295

[24] Brown A. A commentary on CSR. Issues in Social and Environmental Accounting. 2007;**1**:160-164

[25] Suttipun M. Triple bottom line reporting in annual reports: A case study of companies listed on the stock exchange of Thailand (SET). Asian Journal of Finance & Accounting. 2012;**4**:69-92

[26] Fox T, Ward H, Howard B. Public sector roles in strengthening corporate social responsibility: A baseline study. The World Bank, corporate social responsibility practice. In: Corporate Responsibility for Environment and Development Programme International Institute for Environment and Development (IIED). 2002

[27] Melo T, Garrido-Morgado A. Corporate reputation: A combination of social responsibility and industry. Corporate Social Responsibility and Environmental Management. 2012;**19**:11-31

[28] Carroll AB, Shabana CM. The business case for corporate social responsibility: A review of concepts, research and practice. International Journal of Management Reviews. 2010:85-105. DOI: 10.1111/j.1468-2370.2009.00275.x

[29] Lee M. A review of the theories of corporate social responsibility: Its evolutionary path and the road ahead. International Journal of Management Reviews. 2008;**10**(1):53-73

[30] Aras G, Crowther D. Governance and sustainability: An investigation into the relationship between corporate governance and corporate sustainability. Management Decision. 2008;**46**(3):433-448

[31] Organization for Economic Co-operation and Development (OECD). Report of the World Commission on Environment and Development: Our Common Future. [Online]. 1987. Available from: http://www.un-documents.net/our-common-future.pdf [Accessed: December 27, 2011]

[32] Christofi A, Christofi P, Sisaye S. Corporate sustainability: Historical development and reporting practices. Management Research Review. 2012;**35**:157-172

[33] Hopwood B, Mellor M, O'Brien G. Sustainable development: Mapping different approaches. Sustainable Development. 2005;**13**:38-52

[34] Williams B, Wilmshurst T, Clift R. Sustainability reporting by local government in Australia: Current and future prospects. Accounting Forum. 2011;**35**:176-186

[35] Moneva J, Archel P, Correa C. GRI and the camouflaging of corporate unsustainability. Accounting Forum. 2006;**30**:121-137

[36] Gasparatos A, El-Haram M, Horner M. The argument against a reductionist approach for measuring sustainable development performance and the need for methodological pluralism. Accounting Forum. 2009;**33**:245-256

[37] Aras G, Crowther D. Making sustainable development sustainable. Management Decision. 2009;**47**(6):975-988

[38] Turner RK. Sustainability auditing and assessment challenges. Building Research & Information. 2006;**34**:197-200

[39] Cairms RD. On accounting for sustainable development and accounting for the environment. Resources Policy. 2006;**31**:211-216

[40] Guthrie J, Farneti F. GRI sustainability reporting by Australian public sector organizations. Public Money & Management. 2008;**28**(6):361-366

[41] Roca LC, Searcy C. An analysis of indicators disclosed in corporate sustainability reports. Journal of Cleaner Production. 2012;**20**:103-118

[42] Gray R. The social accounting project and accounting organizations and society: Privileging engagement, imaginings, new accountings and pragmatism over critique? Accounting Organizations and Society. 2002;**27**:687-707

[43] Gray, R, Milne, MJ. Sustainability Reporting: Who's Kidding Whom? Accountability Transparency Sustainability. [Online]. 2002. Available from: http://sta-res.st-andrews.ac.uk/media/csear/discussion-papers/CSEAR_dps-sustain-whoskidding.pdf [Accessed: May 5, 2015]

[44] Walter GR, Wilkerson O. Community sustainability auditing. Journal of Environmental Planning and Management. 1998;**41**:673-691

[45] Marrewijk MV. Concepts and definitions of CSR and corporate sustainability: Between agency and communion. Journal of Business Ethics. 2003;**44**:95-105

[46] Bell J, Soybel V, Turner R. Integrating sustainability into corporate DNA. Journal of Corporate Accounting & Finance. 2012:71-82. DOI: 10.1002/jcaf.21755

[47] Gao S, Zhang J. Stakeholder engagement, social auditing and corporate sustainability. Business Process Management Journal. 2006;**12**:722-740

[48] Daub C-H. Assessing the quality of sustainability reporting: An alternative methodological approach. Journal of Cleaner Production. 2007;**15**:75-85

[49] Sciulli N. Sustainability reporting by local councils in coastal regions: An Australian study. Asian Journal of Finance & Accounting. 2009;**1**(1):76-86

[50] Gray R, Owen D, Adams CA. Accounting and Accountability: Changes and Challenges in Corporate Social and Environmental Reporting. London: Prentice Hall; 1996

[51] Criado-Jiménez I, Fernández-Chulián M, Husillos-Carqués F, Larrinaga-González C. Compliance with mandatory environmental reporting in financial statements: The case of Spain (2001-2003). Journal of Business Ethics. 2008;**79**:245-262

[52] Bouten L, Everaert P, Liedekerkeb L, Moord L, Christiaens J. Corporate social responsibility reporting: A comprehensive picture? Accounting Forum. 2011;**35**:187-204

[53] KPMG. The KPMG survey of corporate responsibility reporting 2013. KPMG International. [Online]. 2013. Available from: https://www.kpmg.com/Global/en/IssuesAndInsights/ArticlesPublications/corporate-responsibility/Documents/corporate-responsibility-reporting-survey-2013-exec-summary.pdf [Accessed: May 5, 2013]

[54] Skouloudis A, Evangelinos KI. Sustainability reporting in Greece: Are we there yet? Environmental Quality Management. 2009:43-60. DOI: 10.1002/tqem.20235

[55] Elkington, J. Cannibals with forks—the triple bottom line of 21st century business. Observatoire du Management Alternatif. Paris. [Online]. 1997. Available from http://appli6.hec.fr/amo/Public/Files/Docs/148_en.pdf [Accessed: July 24, 2014]

[56] Elkington J. Governance for sustainability. Corporate Governance. 2006;**14**(6):522-530

[57] Milne MJ, Gray R. W(h)ither ecology? The triple bottom line, the global reporting initiative, and corporate sustainability reporting. Journal Business Ethics. 2013;**118**:13-29

[58] Owen D. Recent developments in European social and environmental reporting and auditing practice—A critical evaluation and tentative prognosis. International Centre for Corporate Social Responsibility. [Online]. 2003. Available from: http://195.130.87.21:8080/dspace/bitstream/123456789/1113/1/03-Recent%20developments%20in%20European%20social%20and%20environmental%20reporting%20and%20auditing%20practic.PDF [Accessed: December 27, 2011]

[59] Haque S, Pathrannarakul P, Phinaitrup B. Modernizing public sector organization: Enhancing coordination and communication by the application of e-government technology. International Journal of Independent Research and Studies. 2012;**1**:135-141

[60] Burritt R, Welch S. Accountability for environmental performance of the Australian Commonwealth public sector. Accounting, Auditing & Accountability Journal. 1997;**10**(4):532-561

[61] Larrinaga-González C, Bebbington J. Accounting change or institutional appropriation? A case study of the implementation of environmental accounting. Critical Perspectives on Accounting. 2001;**12**:269-292

[62] Ball A. A sustainability accounting project for the UK local government sector? Testing the social theory mapping process and locating a frame of reference. Critical Perspectives on Accounting. 2004;**15**(8):1009-1035

[63] Ball A. Environmental accounting and change in UK local government. Accounting, Auditing & Accountability Journal. 2005;**18**(3):346-373

[64] Ball A, Grubnic S. Sustainability accounting in the public sector. In: Unerman J, Bebbington J, O'Dwyer B, editors. Sustainability Accounting and Accountability. London: Routledge; 2007. pp. 243-265

[65] Ball A, Bebbington J. Editorial: Accounting and reporting for sustainable development in public service organizations. Public Money & Management. 2008;**28**(6):323-326

[66] Broadbent J, Guthrie J. Public sector to public services: 20 years of "contextual" accounting research. Accounting, Auditing & Accountability Journal. 2008;**21**(2):129-169

[67] Larrinaga-González C, Pérez-Chamorro V. Sustainability accounting and accountability in public water companies. Public Money & Management. 2008;**28**(6):337-343

[68] Lewis T. Debate: Public sector sustainability reporting—Implications for accountants. Public Money & Management. 2008;**28**(6):329-331

[69] Burritt R, Schaltegger S. Sustainability accounting and reporting: Fad or trend? Accounting, Auditing & Accountability Journal. 2010;**23**(7):829-846

[70] Sciulli N. Influences on sustainability reporting within local government. International Review of Business Research Papers. 2011;**7**(2):282-291

[71] Gray R, Laughlin R. It was 20 years ago today Sgt Pepper: Green accounting and the Blue Meanies. Accounting, Auditing & Accountability Journal. 2012;**25**:228-255

[72] Decreto-Lei 133/2013, 3 outubro 2013. Diário da República, 1ª série – N° 191. [Online]. Available from: http://www.pgdlisboa.pt/leis/lei_mostra_articulado.php?nid=1992&tabela=leis [Accessed: February 26, 2014]

[73] Brown HS, Jong M, Lessidrenska T. The rise of the global reporting initiative: A case of institutional entrepreneurship. Environmental Politics. 2009;**18**(2):182-200

[74] Toppinen A, Li N, Tuppura A, Xiong Y. Corporate responsibility and strategic groups in the forest based industry: Exploratory analysis based on the global reporting initiative (GRI) framework. Corporate Social Responsibility and Environmental Management. 2012;**19**:191-205

[75] Global Reporting Initiative (GRI). Diretrizes para a Elaboração de Relatórios de Sustentabilidade (DERS). Versão 3.0, em português. [Online]. 2006. Available from: https://www.globalreporting.org/resourcelibrary/Portuguese-G3-Reporting-Guidelines.pdf [Accessed: February 28, 2012]

[76] Global Reporting Initiative (GRI). Pontos de Partida – Relatórios de Sustentabilidade da GRI: Quanto vale essa jornada? [Online]. 2013

[77] Willis W. The role of the global reporting Initiative's sustainability reporting guidelines in the social screening of investments. Journal of Business Ethics. 2003;**43**:233-237

[78] Dingwerth K, Eichinger M. Tamed transparency: How information disclosure under the global reporting initiative fails to empower. Global Environmental Politics. 2010;**10**:74-96

[79] Ball A, Broadbent J, Jarvis T. Waste management, the challenges of the PFI and "sustainability reporting". Business Strategy and the Environment. 2006;**15**:258-274

[80] Gray R. Is accounting for sustainability actually accounting for sustainability…and how would we know? An exploration of narratives of organizations and the planet. Accounting, Organizations and Society. 2010;**35**:47-62

[81] Elkington J. Partnerships from CmMs mth forks: The triple bottom line of 21st-century business. Environmental Quality Management. 1998;**8**(1):37-51

[82] Elkington, J. Chapter 1. Enter the triple bottom line. [Online]. 2004. Available from: http://kmhassociates.ca/resources/1/Triple%20Bottom%20Line%20a%20history%201961-2001.pdf [Accessed: July 24, 2014]

[83] Lamberton G. Sustainability accounting—A brief history and conceptual framework. Accounting Forum. 2005;**29**:7-26

[84] Antoni M, Hurt Q. Applying the global reporting initiative (GRI) for public bodies in the South African context: The eThekwini experience. Development Southern Africa. 2006;**23**(2):251-263

[85] Comyns B, Figge F, Hahn T, Barkemeyer R. Sustainability reporting: The role of "search," "experience" and "credence" information. Accounting Forum. 2013;**3**:231-243

[86] Lynch B. An examination of environmental reporting by Australian state government departments. Accounting Forum. 2010;**34**:32-45

[87] Global Reporting Initiative (GRI). G4 sustainability reporting guidelines. [Online]. 2013. Available from: https://www.globalreporting.org/standards/g4/pages/default.aspx [Accessed: February 2, 2016]

[88] Farneti F, Guthrie J. Sustainability reporting by Australian public sector organizations: Why they report. Accounting, Auditing and Accountability. 2009;**33**:89-98

[89] Dumay J, Guthrie J, Farneti F. GRI sustainability reporting guidelines for public and third sector organizations. Public Management Review. 2010;**12**:531-548

[90] Legendre S, Coderre F. Determinants of GRI G3 application levels: The case of the fortune global 500. Corporate Social Responsibility and Environmental Management. 2012:1-11. DOI: 10.1002/csr.1285

[91] Milne M, Tregidga H, Walton S. Words not actions! The ideological role of sustainable development reporting. Accounting, Auditing & Accountability Journal. 2009; **22**(8):1211-1257

[92] Prado-Lorenzo J-M, Gallego-Alvarez I, Garcia-Sanchez IM. Stakeholder engagement and corporate social responsibility reporting: The ownership structure effect. Corporate Social Responsibility and Environmental Management. 2009;**16**:94-107

[93] Global Reporting Initiative (GRI). Sector supplement for public agencies, pilot version 1.0. [Online]. 2005. Available from: www.globalreporting.org [Accessed: May 5, 2012]

[94] Global Reporting Initiative (GRI). GRI reporting in government agencies. [Online]. 2010. Available from: www.globalreporting.org [Accessed: May 5, 2012]

[95] Grindle M, Hilderbrand M. Building sustainable capacity in the public sector: What can be done? Public Administration and Development. 1995;**15**:441-463

[96] Lamprinidi S, Kubo N. Debate: The global reporting initiative and public agencies. Public Money & Management. 2008;**28**:326-329

[97] Mio C. Corporate social reporting in Italian multi-utility companies: An empirical analysis. Corporate Social Responsibility and Environmental Management. 2010;**17**:247-271

[98] Boiral O. Sustainability reports as simulacra? A counter-account of A and A+ GRI reports. Accounting, Auditing & Accountability Journal. 2013;**26**(7):1036-1071

[99] Fernandez-Feijoo B, Romero S, Ruiz S. Effect of stakeholders' pressure on transparency of sustainability reports within the GRI framework. Journal of Business Ethics. 2014;**122**:53-63

[100] Boiral O, Henri J-F. Is sustainability performance comparable? A study of GRI reports of mining organizations. Business & Society. 2015;**56**(2):283-317

[101] Godha A, Jain P. Sustainability reporting trend in Indian companies as per GRI framework: A comparative study. South Asian Journal of Business and Management. 2015;**4**(1):62-73

[102] Denčić-Mihajlov K, Zeranski S. Development of sustainability indicators: Approaches, challenges and opportunities. Facta Universitatis Series: Economics and Organization. 2017;**14**(4):291-306

[103] Ioannou, I. Serafeim, G. The Consequences of Mandatory Corporate Sustainability Reporting, Harvard Business School, 2017. pp. 1-49. [Paper Online]. Available from: http://www.hbs.edu/faculty/Publication%20Files/11-100_ed78b358-dddd-41f0-9a05-5c1b430b15f9.pdf [Accessed: February 5, 2018]

[104] Talbot D, Boiral O. GHG reporting and impression management: An assessment of sustainability reports from the energy sector. Journal of Business Ethics. 2018;**147**(2):367-383

[105] Global Reporting Initiative (GRI). [Online]. 2013. Available from: https://www.globalreporting.org/information/news-and-press-center/Pages/Application-Levels-all-you-need-to-know.aspx [Accessed: May 5, 2013]

[106] Global Reporting Initiative (GRI). GRI application levels (2000-2011), version 3.0. 2011. [Online]. Available from: https://www.globalreporting.org/resourcelibrary/G3-Application-Levels.pdf [Accessed: May 6, 2013]

[107] Yin R. Case Study Research: Design and Methods. 3rd ed. Thousand Oaks, California: Sage Publications; 2003

[108] Zainal, Z. Case study as a research method. Jurnal Kemanusiaan. 2007;**9**:1-6. [Online]. Available from http://www.management.utm.my/jurnal-kemanusiaan/attachments/article/163/JK9_163.pdf [Accessed: April 2, 2016]

[109] Brignall S, Modell S. An institutional perspective on performance measurement and management in the "new public sector". Management Accounting Research. 2000; **11**:281-306

[110] KPMG. International Survey of Corporate Responsibility Reporting 2011. 2011. [Online]. Available from: https://www.kpmg.com/PT/pt/IssuesAndInsights/Documents/corporate-responsibility2011.pdf [Accessed: May 6, 2012]

[111] Rodrigues LL, Tejedo-Romero F, Craig R. Corporate governance and intellectual capital reporting in a period of financial crisis: Evidence from Portugal. International Journal of Disclosure and Governance. 2016:1-29. DOI: 10.1057/jdg.2015.20

[112] Giannarakis G, Theotokas I. The effect of financial crisis in corporate social responsibility performance. International Journal of Marketing Studies. 2011;**3**(1):2-10

[113] Dias A, Rodrigues LL, Craig R. Global financial crisis and corporate social responsibility disclosure. Social Responsibility Journal. 2016;**12**(4):654-671

[114] Cerin P, Scholtens B. Editorial linking responsible investments to societal influence: Motives, assessments and risks. Sustainable Development. 2011;**19**:71-76

[115] Brown HS, Jong M, Levy DL. Building institutions based on information disclosure: Lessons from GRI's sustainability reporting. Journal of Cleaner Production. 2009;**17**:571-580

Chapter 8

Communication Society and Security: Current Threats and Legal Maintenance

Anna A. Chebotareva, Vladimir E. Chebotarev and Alexander S. Rozanov

Additional information is available at the end of the chapter

http://dx.doi.org/10.5772/intechopen.75756

Abstract

Over many centuries, human societies across the globe have established progressively closer contacts. Recently, the pace of globalization has dramatically increased. Unprecedented changes in communications, transportation, and computer technology have given the process new impetus and made the world more interdependent than ever. Information resources and structures have become a tool for achieving a strategic advantage. The authenticity, credibility, and an adequate reflection of information realities represent the key challenges for the communication society. Our research aims to analyze the possibilities of establishing a profound system for countering legalization of proceeds from crime (money laundering) and creating efficient barriers against cybercrimes, such as hacking of personal data. The sphere of security of online communication processes has become an objective element of our life, and it couldn't be ignored, especially due to further development of securing biometric personal data mechanisms.

Keywords: communication society, communication security, terrorism, offenses in the communication sphere, mechanism of remote authentication and identification, biometric personal data, money laundering, legalization of proceeds from crime

1. Introduction

The issue of balance of interests in the so-called triad "state-society-person [1, 3]" has traditionally been considered and settled by representatives of various humanitarian sciences; however, this problem is getting more and more relevant within the global communication society.

Pointing to the dialectical relationship as a characteristic of many components of the information society in the era of globalization, the authors of the research investigate the factors manifested in threats and challenges to security, the issues of reliability and objectivity of information circulated in large volumes, and the impossibility of reliable identification of online communication relations' subjects. The initiatives of the Russian parliamentarians on the formation of remote authentication and identification mechanisms, in particular, related to the creation of the national biometric platform are carefully analyzed. The resonance of the creation of national biometric platform is connected with the amendment to the law, according to which banks will be obliged to transfer biometric data of the clients to the Ministry of Internal Affairs and the Federal Security Service of the Russian Federation (FSS) for the sake of ensuring national defense, state security, law enforcement, and efficient counter-terrorism mechanisms.

Every regime of law is comprised of two essential parts: a coercive, or adjudicative, aspect and a persuasive, or educative, aspect. In the short term, a mode of governance could impose order by shear force [14]. But, over time, establishing an atmosphere of stability and continuity requires the public to understand legal authority in terms of the benefits it confers, and they must be taught habits of compliance. This combination of methods for bringing order to human life and shape to human thought forms the basis of a legal culture. In this regard, the following question seems crucially important: how to preserve the natural rights and freedoms of the individual in order to build effective communications within the society [12]? The answer is rather complicated and controversial. The process of global virtualization has made the society even more open and dependent on information data on the one hand but turned it into a highly vulnerable structure on the other hand.

The genre of our research could be characterized as a mixture of both types—an essay and an overview of the latest legislative initiatives. In this regard, the methodological basis of the research is a complex of general scientific and special legal methods of cognition. When analyzing the regularities of development of a person within a global communication society, we have been applying a dialectical approach, which has allowed us to conduct a study on the basis of two key principles—the principle of determinism and the unity of historical and logical elements of social development. Formal and logical method enabled the authors of the research to analyze the norms of information sphere's legislation, determining the content of the main concepts, and systematize the material in order to obtain generalizing conclusions.

The comparative legal method has contributed to the identification of current trends in communication society's development and to the revision of bibliography on the issue of interaction between communication technologies and society. The system analysis has made it possible to evaluate the existing approaches to legal support of communication security of a person and to relate them to objectively developing social relations. The use of a sociological method helped us to evaluate the factors influencing the behavior of a person as a subject of communication relations. The prognostic approach (in the form of modeling) was used to determine the future prospects of development of the legislation aimed at creating a system of effective legal support of a person's communication security.

The methodology of our research is also based on factorial and causal analysis aimed at identifying the circumstances, which constitute threats to the individual in the communication sphere, and determining the nature of their impact on the efficiency of personal interests' implementation within the IT-space. Considering that factorial analysis as a research tool is used in various fields of knowledge, the authors propose to use this method for a comprehensive and systematic study of the nature of the communication security threats and their influence on personal data security.

2. State, technologies, and legal culture

During the early twentieth century, the nation-state had reached the height of its development and, in its various forms, had come to include within its iterations nearly the entire habitable surface of the earth. It was explicitly defined in its dimensions, protected by recognized borders, and entitled to self-defense according to defined rules of war. It was sovereign in its domestic policies, wielding exclusive authority over its people and resources. Each state was recognized as a member of the family of nations, able to enter into relations with any one of its counterparts as an equal polity.

One reason for the successful proliferation of the nation-state as a form of governance had been that the stage of technical advancement was well adapted to its limited territorial oversight. During the twentieth century, the state had been grounded in well-tested doctrines and practice, but it also fits the level of technical advance prevailing at that time. This included not only the printed book and journal but also railroad and industrial machines and, finally, radio and cinema. Combined together, these were able to create a total environment of public understanding and national purpose within a region of common language and custom.

Difficulties with the mechanism of the state began to emerge in the late twentieth century, however, because of the way further innovations in communication and transportation were being employed. Suddenly, capital could be assembled, labor concentrated, and resources marshaled, without regard to distance or topography. Sound and image could be broadcast across borders and around the world. Information of any quantity on any topic could be transmitted from any one location to any other location at any time, by any person.

For the state, these new developments—not to mention advances in warfare technology—marked a dramatic challenge not only to its functioning as a framework of authority and its foundation of national law but also to its self-sufficiency as a nation. Among the first problems to be confronted was the effective negation of its borders as a protection against unregulated communication and trade. Any former conception of the border as an absolute and defined barrier separating not only territories but also legal jurisdictions was becoming untenable.

With the great concession of the border, the next problem for national governments was the regulation of affairs that involved immigrants, visitors, and commercial agents who had entered from outside. The problem was how, on a practical level, these could be regulated or even monitored. Overseeing the affairs of its own citizens was rather easy, because both they and their property generally existed within the region defined by national borders. But matters were less simple with those entities whose primary assets and ownership lay outside territorial limits and beyond the reach of authority.

Moreover, the new technical innovations brought instability and imbalance on an international scale as well (for instance, an outbreak of Third World protest and resistance, against what was seen as the Western exploitation of foreign and disadvantaged countries).

Complicating this reaction was another unforeseen factor: television. With the new ability to transmit sound and image, messages of discontent and revolution could be widely broadcast. Nationalist, socialist, and anti-capitalist movements in smaller countries—in the past, wholly isolated from the developed centers of power—could garner a sympathetic following around the world. Official explanations of world affairs now seemed to be undermined by the new broadcast media. The result was disruption not only in matters of foreign relations and foreign policy but also in the atmosphere of consensus necessary for quiet order and international stability. These worldwide upheavals culminated in the student rebellions of 1968.

On other hand, the positive impact of those technical innovations should not be underestimated. The arrival of the twenty-first century marked not only the advent of a new millennium but also the onset of a new age. It was termed the age of technology, of communication society, and of globalization. It was sometimes referred to as the postmodern age to distinguish it from the period of modernity that preceded it. In fact, because of remarkable advances in technology, many conventional forms of governance and rule were coming to be reconsidered.

At the beginning of the new millennium, the problems roiling the state were offset by the exhilarating impact the new technologies were having on the corporation. For that structure, electronic transmission of sound and image, voice communication, online communication systems, and computerization were unqualified benefits. The new ability to travel, to transport, and to trade opened vistas of opportunity for expansion and consolidation.

In this regard, *Andrew Feenberg*'s concept of technology (Andrew Feenberg is a proponent of a dialectical theory of technology) provides a rather clear understanding of an ambivalent nature of the interaction between the modern society and technologies: "Technological development is overdetermined by both technical and social criteria of progress, and can therefore branch in any of several different directions depending on the prevailing hegemony. … While social institutions adapt to technological development, the process of adaptation is reciprocal, and technology changes in response to the conditions in which it finds itself as much as it influences them" [6].

Technical advances greatly strengthened the ability to manage and control, extending commercial opportunities to the most remote regions of the earth. It gave great impetus to international business firms, accelerating even more the proliferation of decentralized multinational corporations. Along with these was the continuing proliferation of worldwide television broadcasting, a medium that brought enormous commercial opportunity in both entertainment and advertising.

Finally, through the medium of television and other channels and networks of communication, the digital sphere could assume an educative function as well. Of course, during the human history and still now, the state has played a crucial role in this process, but with the new advances; nothing could equal the various electronic media as instruments for shaping public culture and consciousness. Unlike the old brick-and-mortar national school system, the new media could, in effect, create an artificial reality that was continuous and ubiquitous in its effects. Nothing could match its potential ability to instill habits of acceptance and compliance, a crucial necessity for an extended rule of law.

The authors of this research furthermore call to pay special attention to the issue of legal support of individuals' communication security, which is constrained by extremely complicated challenges and threats, including:

- The growth of offenses in the sphere of virtual space
- The legal issues of identification and authentication of the individual
- The imperfection of applied information and telecommunication technologies, including the possibility of technological failures, and a vulnerability of personal databases in e-government and e-justice systems
- The preservation of digital inequality
- The imperfection of a digital economy

Meanwhile, we propose to consider the communication security of a person as a state of his/her personal safety, which is determined by minimization of risks (in the form of internal and external information threats) for the individual in the conditions of global communication society, the ability to confront these challenges on the basis of a culture of communication security, as well as the formation of a state policy aimed at creation of conditions for the efficient implementation of a person's rights and freedoms in the sphere of global communications.

The interests of the individual in the online communication sphere include the implementation of constitutionally enshrined human/citizen's rights to access to information, including the use of information for the purposes of non-prohibited activities (physical, spiritual, and intellectual development), and also the protection of information that provides personal security. In this sense, we absolutely agree with *Fuchs*'s perception of the information as a dialectical process that establishes an interconnection of subjects and objects via a threefold process of cognition, communication, and cooperation [7].

It is appropriate to highlight the following legal culture mechanisms and means of providing personal communication security to be improved in the near future:

- The mechanisms aimed at the identification and authentication of individuals in the communication space, including means of biometric identification, mechanisms to ensure the protection of information, its integrity and reliability, as well as technologies of personal data protection.

- There is also a need for real adoption of interlinked organizational, scientific and technical, information and analytical, economic, and other measures to anticipate, detect, deter, prevent, and fend the information threats, as well as to eliminate the consequences of their manifestation.
- From the Russian perspective, the most urgent need of a present moment is to address the issues of high technologies development (following the example of innovative, universal blockchain platform, a quantum cryptography, etc.), the implementation of the state policies in the field of import substitution, the development of the digital economy, and the issue of a legal regulation of cryptocurrency.

The blockchain technology represents a decentralized distributed database of all confirmed transactions, whose functioning is based on cryptographic algorithms. Such technology allows:

a. To record reliable data on ownership of an asset, existed in digital format, to a certain person without the need to attract any specialized intermediary (in this case, it can be considered as a strong factor in the disintermediation of the economy)
b. To ensure that such an asset can be directly transferred to another person

In our opinion, the other relevant challenge to be resolved as soon as possible is the development of communication security's quantum cryptography system. The issues, linked to legal provision of personal communication security, are directly connected with the development of the digital currency world market (including such e-financial instruments as Bitcoin, Ethereum, etc.).

The existence of cryptocurrency transactions' anonymity without a legal regulation could be considered as an absolute risk factor, because this sort of operations is decentralized, and the absence of intermediaries leads to a quiet understandable interest of criminal organizations to gain huge profits by committing cybercrimes (including online theft and hacking). In this regard, it is necessary to officially recognize the cryptocurrency and start the legal regulation of this sphere for the purpose of ensuring communication (and also digital) security. Legal regulation of crypto-space can't be ignored; this is an objective necessity.

3. Social aspect of globalization and communication skills

Has globalization made the mankind smaller? Has the world become more intimately connected? Are the boundaries between global and local and public and personal eroding? Is this the time when nobody is perfect and no doctrine is universal? Is this the time when the abundance of choice leads to a decision paralysis? Is this the time when the world economic leaders need a facilitator to agree on a strategy for the global economy?

It has become clear that to live in a more complex, interconnected, and detailed world, the mankind needs to be equipped with tools for understanding its own behavioral patterns and their impact on the new social dynamics. Survival in the new world means being able to navigate a gradually more unstructured and rich in detail social universe, being able to sort across a greater array of options, and possessing a deeper and more strategic understanding of self with lesser time for learning and self-discovery. What are the emerging solutions to the challenges of survival in a globalized world?

In recent years, there has been an explosion of research in psychology and social studies aimed at gaining an insight into what motivates us as species, what makes us change our ways, and what makes us happy [8, 11]. A new body of research focused on gender dynamics has provided us with a greater awareness of the needs of each gender. It explained the reasons that underlie gender-based social protocols and gave many the tools for improving their interactions with the other gender. Studies of consumer behavior have made us aware of how our brains and our feelings work when exposed to different external stimuli.

From the conversations we have to the words we choose, to the questions we ask, and to the way we think, each action or inaction speaks volumes about us. Understanding these clues and noticing the differences among individuals at an early age will make us more receptive, aware, influential, and evolved as species.

Despite the abundance of social research and the growing necessity to know the techniques for effective social interactions early in life, this valuable information has been slow in getting into school or university curriculums. Social studies continue to be confined to auxiliary subjects in business schools and catchall in university curriculums or remain in the form of bestsellers and articles in popular psychology magazines.

High schools and universities should integrate social studies and communications into a discrete social skills discipline and teach it in the way a foreign language is taught.

More scientific research of social networks is also needed. Social networks are the most vivid manifestations of globalization. They have already shown their capabilities in changing social structures, bringing people together, and creating greater prosperity in underdeveloped locales. Social networks if partnered with (and not just observed) can inform a variety of disciplines, from history and sociology to business and management. The IT industry and academia must join forces to produce new analytical data on social tools that facilitate growth and development of mankind, scale successes, and improve social climate.

4. Globalization, communication society, and Russian legal culture

Undoubtedly, the twenty-first century can be considered as the era of information accompanied by the global computerization of the modern society. The current dependence of human civilization on communication component has made it much more vulnerable in this regard. Moreover, such vulnerability is linked to the fact that our society represents itself a fundamentally open structure. The most important result of the global communication society's formation is the emergence of the so-called global communication space.

The process of society's computerization is making major changes in the social structure and the existing mechanisms for social decision-making process at all levels [17]. Globalization of social relations leads to their transformation, including new mechanisms of control and decision-making processes. Getting knowledge of ongoing social processes, discussion of trends, and forecasting possible outcomes are an important part of the goal of achieving sustainable development [7]. Nowadays the role of social consolidation of the population in the countries and regions has become even more important. Human potential in favorable conditions for the development of freethinking, informed, and responsible person could become a factor of stability, as well as successfully deal with new challenges in the form of extremism, racism, intolerance, and moral degradation [15].

Globalization of communication society is a macroscale, multifaceted, and internally contradictory process of the growth of similarity in the world communication systems (economic, political, social, and legal).

Many processes in the communication society are in a dialectical relationship and interdependence, and these relations are complex and contradictory. According to the authors of this study, the use of historical and dialectical approach (a historical context of communication and the unity of historical and logical elements of social interactions have been carefully studied by *Putnam and Pacanowsky* in 1983; for more details, see [13]) for analyzing the phenomenon of communication society is due to a number of factors [3, 4]:

1. A certain degree of inertia of the communication society and its unwillingness to fully perceive the products of scientific and technological progress (due both to objective and subjective reasons).
2. While developing and improving, the information society is not going in the direction of reducing all kinds of threats, but, on the contrary, both the number and intensity of such threats are constantly increasing [5]. The information environment is in constant development, it is moving, it is not static, and—as a result—such environment is facing obvious vulnerabilities and risks
3. The excessive amount of information is increasing exponentially; such situation leads to the fact that a person is not prepared to perceive it. As a result, the so-called internal filters have appeared: people automatically "filter" the information even before its perception, highlighting only necessary and important data for themselves.
4. The parallel coexistence of two trends: a formation of large amount of databases ("big data") with general information and—at the same time—a lack of relevant and useful information (*Ronald Day*, for instance, indicates that recently the information and communication products have been treated mainly as "reified and commoditized notion"; for more details, see [4]). The issue of reliability of the received information has become even more urgent. Another dangerous threat deserves special attention—the overload of information flows with harmful and prohibited data, as well as the misinformation.
5. The irregular and unbalanced character of information technology implementation (for comparison: in contrast to electronic workflow, paper workflow has been evolving over

the centuries). As a consequence, we are facing the mistrust to the process of implementing e-government, as well as to providing public and municipal services in an electronic form.

6. Digital technologies, used for process automation, do not have a complete form; they are in a constant process of improvement and replacement with latest upgrades. Hence, the existing solutions in the field of process automation are perceived as temporary.

7. Communication society in the context of globalization, based on cross-border concept (in this regard, *Judith Martin*'s concept of intercultural communication has a crucial meaning; see [10]), elevates the anonymity in the networks and, in turn, the identification of subjects of information relations to a level of a fundamental problem.

The central subject of communication relations [18]—a person or an identity—is subjected to serious challenges and threats; hence, the state of its security needs special attention.

The most important attributes of the personality in the modern global information world include a set of personal information, which cannot be reliably protected by technical and software means only. Personal information of the individual will be inevitably accumulated and fixed in the Internet environment (or digital space). It can be distorted and supplemented by false information that will harm the individual in terms of his or her reputation, image, breach of secrecy, etc. The person in the modern world is deprived of local protection in a macroscale environment with no national, linguistic, cultural, and even ethical boundaries.

The shift of interpersonal communication, as well as communication with society and the state into the environment, generated by the developing information and telecommunication technologies, creates conditions for a high vulnerability of the personality in the global world.

The interests of a person in the information sphere are to meet all his/her possible needs—to ensure the right to access to information; the possibility of citizen's participation in lawmaking activities, including through the development of electronic democracy mechanisms; the possibility of obtaining state and municipal services in electronic form, as well as the implementation of the right to protection through electronic justice mechanisms; etc.

The global information society appears as a platform for the development of both positive and deterrent factors: the first one contributes to the realization of the whole spectrum of interests of the individual; the last one hinders the development of the information society itself (as a whole).

The global, cross-border nature of information and telecommunication technologies leads to the immensity of offenses in the information sphere; this trend, in turn, leads to a significant violation of the rights and interests of the individual. The significance of risks and threats can cause serious damage in the implementation of personal interests in the global communication society. This fact can be proved by:

- The identification issues
- The possibility of falsifying the results of online voting ("e-voting" [14])
- The possibilities of technological failures in the process of development of electronic parliament elements and electronic democracy mechanisms [9]

- The possibility of unreliability of databases
- The insecurity of confidential information and personal data in the provision of public and municipal services in an electronic form
- The potential danger of unfair use of personal data in the process of development of electronic justice mechanisms
- The problem of distribution of illegal and harmful content, directly threatening human health, and disorienting the person
- The problem of distribution of defamatory materials by electronic media [18]
- The possibility of theft of information, used in the Internet banking systems [16]
- The possibility of loss of data as a result of malicious attacks while working on the Internet (hacking)

The uniqueness of the virtual environment forces the subjects of information relations to adapt, looking for ways and opportunities for existing in "real-life [3]" conditions. At the moment, for instance, we are facing the processes of transferring credit institutions' activities into the virtual environment and the development of digital financial services. The Russian State Duma (the lower house of the Russian Parliament) has been consistently implementing since April 2017 an initiative to create a legal framework for the use of remote authentication and identification mechanism, through which credit institutions will be able to open accounts to individuals (natural persons) via the Internet. The Project Law "On Amendments to the Federal Law" "About counteraction of legalization (washing) of income gained in the criminal way and to terrorism financing" also regulates the procedure for collection and transfer of personal data (including biometrics) into a single system—a Unified System of Identification and Authentication (USIA). USIA is a system created and developed by the Russian Ministry of Communications within the e-government infrastructure in order to streamline and centralize the processes of registration, identification, authentication, and authorization of users.

According to the plan of the Russian parliamentarians, a citizen will need to come to the bank once in order to provide personal data; soon after that, he or she will be able to open accounts without personal presence. At the same time, the main condition to be satisfied by the remote identification of the customer of the credit institution is the absence of his/her involvement in the legalization and laundering of proceeds from crime, as well as in extremist or terrorist activities. And it is not just about personal data. This initiative supports the procedure of interactive remote authentication and identification of the client of credit institution by using citizen's biometrics.

The document provides for the obligation of banks, included in the special list established by the Central Bank of the Russian Federation, on behalf and with the consent of the client (natural person) in order to conduct further remote identification, including other credit institutions, to collect and transfer the following data into the USIA:

- Last name and first name of a person, person's patronymic (if the other does not follow from the law or national custom), a citizenship, a date of birth, an identity document number, a migration card number, the data of a document confirming the foreign citizen or the person without citizenship on stay (residence) in the Russian Federation, the address of residence (registration) or place of stay, taxpayer identification number (if any), an insurance number of individual personal account, and the number of mobile telephone communications subscriber (mobile phone number).

- Information on client's biometric personal data.
- Information on the client's consent to the processing of his/her personal data, including the unified identification and authentication system and information technology elements that ensure the collection, processing, storage, and provision of biometric personal data.

A range of practical important issues has been revealed during the adoption stage of this draft law (a bill). First of all, the agenda includes the establishment of the system of protection of customers' biometrics, the cost of its implementation, as well as the mechanism or the process of customer's face and voice identification.

At the final stage of drafting of this bill, the State Duma of the Russian Federation had adopted the law on the creation of a mechanism for interactive remote authentication and identification of the customer of the credit institution—a law "On amendments to certain legislative acts of the Russian Federation [3]." This change in the Russian legislation will allow banks to open accounts to individuals (natural persons) without their personal presence, only with the use of biometric passports and data, uploaded to the "Public Services" web portal.

In turn, the law on combating legalization (laundering) of proceeds from crime is supplemented by provisions according to which the banks, included in the list of the Central Bank, on behalf and with the consent of the customer (natural person), can operate with his/her personal data, uploaded into the USIA database.

In general, the system of remote identification of credit institutions' customers (natural persons) is proposed to be based on the processing and use of biometric personal data, since such identification has the highest degree of reliability in the digital space.

The pilot project of the mechanism of implementation of the law is expected to be tested on a limited number of bank operations.

The new law has caused a wide resonance; its reason has appeared during the second reading of the bill. This is an amendment according to which banks will be obliged to transfer biometric data of clients to the Russian Ministry of Internal Affairs and the Federal Security Service (FSS) for the sake of national defense, state security, law enforcement, and counter-terrorism. It is adopted that the order of data transfer will be established by the government. At the same time, it is not indicated that the consent of the bank's client is necessary.

The alleged violation of privacy (Article 23 of the Russian Constitution) and the possibility of issuing personal data to the Ministry of Internal Affairs and FSS have evoked a protest mood. We would like to remind that according to Article 23 of the Constitution of the Russian Federation, everyone has the right to inviolability of private life, a personal and family secret, and protection of the honor and reputation.

There is a certain contradiction of the analyzed norms to the Federal Law "On personal data," which directly prohibits the processing of personal data for purposes not specified in their collection. According to Dmitry Yanin, the President of the International Confederation of Consumer Societies, "... getting the biometric data in exchange for online access to services is an unequal fee, there is a high probability of leakage... In fact, you can consider that biometrics will soon be available to all [15]." In Russia, the problem of availability of financial services is not acute; it is better to go to the bank and not share data—it is difficult to predict who will use them and in what way.

As for the problematic issue of creating a full-fledged protection of biometric data of banks' clients, the law provides that for the provision of biometric personal data of an individual through the channels of information transfer for the purpose of his/her identification without personal presence via the Internet, the encryption (cryptographic) means should be used to ensure the security of transmitted data from security threats, relevant in the processing of biometric personal data [5, 15].

The envisaged obtaining the consent of a citizen of the Russian Federation to the processing of personal data and biometric personal data for the implementation of his/her identification may be signed by his/her simple electronic signature, the key of which is obtained in accordance with the rules for the use of a simple electronic signature when applying for state (public) and municipal services in electronic form, established by the Government of the Russian Federation.

The digital identification operator for the banking sector will be "Rostelecom" (Russian state universal telecom operator), which will create the so-called National Biometric Platform (NBP). At the same time, it is planned to use NBP in medicine, education, and retail, in multifunctional and certification centers, and in departments of the Ministry of Internal Affairs.

NBP is supposed to represent a set of specialized information and technological elements that enable the collection, processing, storage, allocation, and compliance of biometric data.

This platform will be located in the secure cloud infrastructure of "Rostelecom," which will be accessed by banks through special communication channels of the system of interdepartmental electronic interaction (SIEI).

In connection with the latest initiatives of the Russian parliamentarians, a number of "painful" points should be noted. The creation of remote authentication and identification mechanisms is aimed, according to the government's plan, at ensuring security and countering the financing of terrorism and the legalization (laundering) of proceeds from crime. It is on one side of the scale. On the other side, we are witnessing the opportunities for violation of our privacy, the danger of incomplete protection of biometric data of bank customers, as well as the threat of its loss, theft, and free access. The obligation, imposed on banks to transmit client biometric data to the Russian Ministry of Internal Affairs and the Federal Security Service for the sake of national defense, state security, law enforcement, and counter-terrorism, may open up wide opportunities for the abuse of such data.

There is a danger to repeat our previous mistakes. Moreover, when studying the peculiarities of crimes in the banking sector, committed using high technology, on the example of data in

June 2016, we have already noted that Russia ranks second in the world rating in the number of information leaks in the financial sector. At the same time, in 73 percent of cases, customers' personal data had been lost or stolen from the Russian banks. As a result, more than 22.5 million personal data records had been leaked to the Internet.

Russian banking system has more than enough problems without the system of remote authentication and identification, which is created and planned to be implemented. Thus, specialists-practitioners in the field of Internet banking and remote banking are right, stating that "the rapid development of Internet technologies does not allow us to predict all the strategic risks… [1, 15]." For instance, Professor Savenkov notes that "many banking institutions underestimate the threat of hacker attacks, without building an adequate system of information security, and thereby create conditions for large-scale theft of funds [15]."

The result is the fact of toughening criminal liability for the crimes, committed in the mass use of payment services in the context of the increased risk of illegal access, destruction, modification, blocking, copying, provision, and dissemination of information, as well as other illegal actions promoted by the development of high technologies.

Speaking about the so-called sensitive information, it should be noted that the attention to the components of this concept—personal information, private life information, as well as personal data, including biometrics—is drawn from the entire world community.

Recently, the National Association of State Chief Information Officers (NASCIO) has published an action plan for a reasonable investment in cybersecurity entitled "Better Data Security Through Classification: A Game Plan for Smart Cybersecurity Investments [2, 3, 5]."

This plan attempts to classify the data by gravity consequences of unauthorized access to them. At the same time, the document explains why the risk-based approach to cybersecurity is the best option for protecting the data of state organizations. Using this approach, the efficiency of operational management increases, the evaluation of the value of information assets becomes more accurate, the ability of hackers to attack these assets is reduced, and the decision-making process is improved.

Data security is always critical for government agencies, so government IT directors have identified data management and data analytics (including data architecture, big data, predictive analytics, etc.) as a priority for 2017.

The classification is defined as "the process of identifying information that needs to be protected from unauthorized access and misuse [3, 5]." Each federal agency should be the competent classification authority for the data and information it collects or uses for the performance of its tasks.

The following classification is proposed:

1. Critical data—the data of critical importance (it is impossible to carry out the most important state functions without them, e.g., the cadastral records or a register of voters)

2. Sensitive information—the information, if it is disclosed or stolen, which may cause damage to a citizen (e.g., tax data or bank statements)

3. Medical data—the information, which includes a significant amount of personal health information that can be used to discriminate a citizen if it falls into the public domain or into the hands of a hacker

4. The information used to identify people (personally identifiable information or PII)—basically it is data collected by financial and similar institutions

It is necessary to emphasize that the possible compromise of PII can lead to identity theft with a variety of consequences, including primarily the theft of money from the accounts of the citizen.

Other important information that does not fit into these categories, but also requires protection, may be available to government agencies.

The data classification, developed by NASCIO, should be taken into serious consideration. The criterion, laid down in its basis (severity of consequences of unauthorized access to data), contributes to a more accurate assessment of risks (challenges and threats) in the information sphere. And today such assessment is very actual and is capable to minimize committed offenses in the conditions of global and transboundary character of rapidly developing information and telecommunication technologies.

5. Conclusions

In today's society, it is impossible to exist without social networks and different Internet technologies. Currently every person, connected to a computer, has to register in at least one social network. Many people do not care about the security of their personal data. But almost any site requires us to enter basic personal information, such as names and the date of birth. Most visitors of the Internet have the same password on all sites, which is a plus for hackers.

Personal data security is a state of personal data security characterized by the ability of users, technical means, and online communication technologies to ensure confidentiality, integrity, and availability of personal data during their processing in personal data communication systems.

Communication security is becoming a key factor in the provision of electronic services. Modern communication services are distinguished by the use of a large amount of sensitive information that needs protection (personal data, payment information, keys, and secrets).

The possibility of creating a National Biometric Platform could provide remote authentication, based on the biometric features of users in any remote service channels: mobile applications, web clients, or points of contact. Such a platform will become not only more reliable and convenient than password or SMS protection; it will also allow to operate a mechanism of remote access to public services. Thus, it could become a tool for improving the efficiency of interaction between the state, business elites, and society.

Among the possible areas of application of NBP are e-government (public services), telemedicine, distance education, e-commerce (including control of remote purchases of medicines), and financial and legal sectors of the economy in terms of transaction confirmation. It is also possible to use the platform for biometric access control to important infrastructure facilities, such as sports stadiums.

Author details

Anna A. Chebotareva, Vladimir E. Chebotarev and Alexander S. Rozanov*

*Address all correspondence to: rozanov-88@list.ru

Russian University of Transport (The Institute of Law), Moscow, Russian Federation

References

[1] Abbate J. Inventing the Internet. Cambridge: MIT Press; 2000

[2] Bradberry T, Greaves J. Emotional Intelligence 2.0. San Diego: Talent Smart; 2009

[3] Chebotareva A. Cybercrime in the banking sector: The main directions of the criminal policy of the Russian Federation. The Criminological Magazine of the Baikal State University of Economics and Law. 2014:140-144

[4] Day RE. The Modern Invention of Information: Discourse, History, and Power. Carbondale: Southern Illinois University Press; 2001

[5] Tardy T. European Security in a Global Context. Internal and External Dynamics. Routledge; 2010

[6] Feenberg A. Transforming Technology: A Critical Theory Revisited. Oxford: Oxford University Press; 2002

[7] Fuchs C. Information and communication technologies & society: A contribution to the critique of the political economy of the internet. European Journal of Communication. 2009;**24**(1):69-87

[8] Gumucio DA. Making Waves: Stories of Participatory Communication for Social Change. New York: Rockefeller Foundation Report; 2001

[9] Grossman LK. The Electronic Republic: Reshaping Democracy in the Information Age. New York: Viking; 1995

[10] Martin JN. Intercultural Communication in Contexts. New York: McGraw-Hill; 2010

[11] Medina J. Brain Rules. Seattle: Pear Press; 2008

[12] Minchenko T. The dynamic model of freedom of conscience in the modern world. European Social Science Journal. 2014:533-537

[13] Putnam L, Pacanowsky M. Communication and Organizations, an Interpretive Approach. Newbury Park: Sage Publications; 1983. 303p

[14] Rhodes R. Understanding Governance. Policy Networks, Governance, Reflexivity and Accountability. Buckingham; 1997

[15] Savenkov A. Criminal policy and the stability of the financial and credit system. Journal of Russian Law. 2016:78-91

[16] Sychev A, Revenkov P, Dudka A. E-Banking Security. Moscow: RK-Laboratory Image; 2016. 212p

[17] Teisman GR. Models for Research into Decision-Making Processes: On Phases, Streams and Decision-Making Rounds. Public Administration; 2000

[18] Acuña BP, editor. The Evolution of Media Communication. InTech; 2017. DOI: 10.5772/6516

Section 2

Socioemotional Skills in Interpersonal Communication

Chapter 9

Online Dispute Resolution

Deepak Verma, Anshu Banwari and Neerja Pande

Additional information is available at the end of the chapter

http://dx.doi.org/10.5772/intechopen.76032

Abstract

The contemporary world belongs to communication where lot of interactions, trading, and business dealings are happening between people, organizations, and business partners. This has led to the sharp rise in disputes and conflicts, and they have become an unavoidable part of our ecosystem. Disputes and conflicts are not only related to online communication but also crops up during face-to-face or offline communication. The disputes/conflicts need to be handled, managed, and resolved timely and in a cost-effective manner. These days, online communication medium is used very effectively for such dispute resolutions. Online dispute resolution is mostly done through video conferencing, email exchanges, instant chat, and interaction through mobile phones. In order to understand the corporate perspective of India related to the online communication and online dispute resolution, interviews of 50 working professionals were conducted. Each of these professionals is at the leadership position and uses online communication more often in their day-to-day professional work. Their views were recorded and interpreted in order to understand practical issues related to online communication and online dispute resolution. In this chapter, we will be discussing the pertinent process of online dispute resolution for both online and offline disputes pertaining to the business environment.

Keywords: online dispute resolution (ODR), process, tools and impact of ODR, corporate perspective of India

1. Introduction

Online dispute resolution (ODR) is the settlement of disputes through online mode of communication/interaction between the disputed parties. Online dispute resolution undertakes disputes that are partially or fully settled over the Internet, having been initiated in cyberspace but with a source outside it (offline) [1]. In the literature and practice, ODR is also called as an alternate dispute resolution (ADR) and Internet dispute resolution (iDR), and these terms are used interchangeably.

Online dispute resolution (ODR) uses alternative dispute resolution processes to resolve a conflict or dispute. Online dispute resolution can be used for disputes arising from an online mode of communication/transaction and also for disputes not involving the Internet, known as "offline disputes." Online dispute resolution is an alternative to the traditional legal system where litigations are filed in court and justice is sought through legal proceedings. Traditional legal system involves a courtroom, lawyers, judges to register, and resolve the dispute.

The parties involved in dispute may use Internet and web-based technology in a variety of ways. Online dispute resolution can be done entirely on the Internet through email, chat, and videoconferencing and if needed parties may also meet in person for face-to-face interaction. Often, a combination of "online" and "offline" (such as face-to-face) methods are used in online dispute resolution.

ODR systems use synchronous as well as asynchronous communication in the process of resolution [2]. Synchronous communication means communication/interaction is happening on a continuous basis between disputed parties and mediator/arbitrator/negotiator (appointed by the dispute resolution agency), for example, videoconferencing and instant chat. However, in case of asynchronous communication, interaction is happening in intervals or in discontinuous manner, for example, email exchanges. Online dispute resolution can involve the parties in mediation, arbitration, and negotiation.

2. Types of online dispute resolution

Online dispute resolution involves mediation, arbitration, and negotiation to enable the dispute resolution.

2.1. Mediation

Mediation is exercised to initiate a broad-based consultation between parties which are involved in conflict [3]. In this process, mediator, who is a neutral, third-party person does the mediation. Mediation is a private process, where the mediator helps the disputed parties to discuss the issue and try to resolve the dispute. The parties who are involved in the conflict have the opportunity to present their point of views, present evidence, and argue for the satisfactory resolution. Mediation can be done entirely online with Internet technology, and parties can have a face-to-face interaction also by meeting personally in the presence of the mediator if required. Some online dispute resolution methods involve a combination of these methods. In the mediation process, the mediator is only a facilitator and does not have the discretion to make a decision. Mediation resolves the dispute while maintaining a consensual and nonbinding approach [4]. Mediator interacts with the parties in dispute to set common rules for the resolution process and assist the parties by identifying and pointing out possibilities for agreement/solution between parties, and it is up to the parties to agree or disagree toward the end of the process. After seeking the parties' consent, the mediator may evaluate the parties' rights/interests and suggest possible settlements. The resolution of the dispute is the choice of involved parties only, and they will have to reach an agreement/settlement. If they will reach an agreement, the parties complete a written documentation that contains the complete specification of the settlement. In many instances, parties go to the court for the enforcement of the agreement.

2.2. Arbitration

Arbitration is a process where a third party makes a decision about the dispute after going through the issues, arguments, and evidence. Arbitration is different from mediation because the neutral arbitrator has the authority to make a decision about the dispute. This neutral arbitrator is approached by the involved parties and seeks his help for the resolution of the dispute. As compared to traditional litigation process, arbitration is less formal, less complicated, and more importantly can be completed in lesser time. Like mediation, this process can also be done entirely online with Internet technology (email, instant chat, or videoconferencing) or the parties can physically meet in the presence of the arbitrator. Arbitration is either binding or nonbinding. In binding arbitration, the parties are supposed to follow the arbitrator's decision. The arbitrator's decision may get the enforcement from the court also. Generally, the arbitrator's decision has an influence on the final judgment of the court. However, in nonbinding arbitration, the arbitrator's decision is only advisory and not binding, and it is up to the parties' discretion to follow the judgment. In case of further proceedings of the dispute in court, the arbitrator's decision may be submitted as evidence. The arbitration process is similar to a court trial, but it has less rules as compared to trial court rules of evidence and the arbitrator need not apply the governing law in totality. After the hearing process, the arbitrator issues a decision mostly along with the required explanation for his opinion. The arbitrator's decision can be enforced by the court if the involved parties have agreed toward a binding arbitration. Only in certain cases, the arbitrator's decision is appealed and overturned by the court.

2.3. Negotiation

Negotiation is defined as a process of making offers, counteroffers, and concessions in between two or more parties so that they can arrive at an agreement [4, 5]. Negotiation can also be taken as an emotional and complex decision-making process which assists in reaching for an agreement in exchange of goods and services [5, 6]. ODR is not much used in case of negotiation, but mostly it is used in mediation and arbitration.

3. Process of online dispute resolution

Dispute resolution through online mode came actually into light in late 1990, when Internet access became easily available and more and more people started using emails and web-based communication. Online dispute resolution evolves not only to resolve disputes arising from online communication but also to resolve offline disputes.

We have both automated and assisted mode of mediation to resolve online and offline disputes. First, we will discuss about assisted online dispute resolution. The ways and means of assisted online dispute resolution are as follows:

In case of assisted online dispute resolution, a mediator coordinates between the parties in dispute and tries to resolve the matter. Here, the mediator mostly uses the email as the communication mode and both the disputed parties access each other's statements. The mediator tries to arrive at the acceptable terms for both the parties and resolve the issue. Other than the email, chat rooms and videoconferencing are also used for instant messaging for dispute resolution. These days due to the advancement in technology and better connectivity, videoconferencing provides lot of convenience. Videoconferencing not only enables live conversation, but it also provides personal connection because of virtual face-to-face interaction. However, usage of email or instant chat or videoconferencing is subject to the nature of the dispute and availability of the persons involved [7].

Disputes originating from offline communication also get resolved with the same process as mentioned earlier. Presently, there are many professional companies that are providing ODR services to their clients. These companies are easily accessible and their service charges are also quite competitive and worthy to their efforts. The mediator, appointed by the company with their constant interaction with the client, acts as a facilitator as well as the evaluator. Sometimes, mediator needs to guide the involved parties and direct them to arrive at an agreement. Hammond [8] conducted a study on the process of online mediation and mediators and pointed out that a mediator needs to adept his communication behavior keeping in mind the online environment. For the mediator, good listening skills and ability to ask relevant and to the point questions are very critical to provide clarity to the involved parties.

3.1. Automated online dispute resolution

Automated dispute resolution is mostly used in cases where dispute has originated because terms have not been agreed/negotiated properly [8]. Here, automated systems help in negotiation between disputed parties and the system usually helps in arriving at a win-win situation for disputants. There are many software packages available, which assist in negotiation through blind-bid negotiation [9]. Here, the disputed parties submit their acceptable amount to them in the online portal system/software, and then the software checks the amount entered in the system and points out whether the deal will materialize between the disputed parties or not. If the amount entered by both the parties matches and overlaps, then the software confirms the deal, and if there is lot of gap between the agreed amount, the deal is called off. Simultaneously, there are softwares, which use game theory concepts to help in resolving disputes. These softwares are quite sophisticated and apply adjusted winner procedure. This adjusted winner procedure was coined by Brams and Taylor [11]. This procedure is based on the assumption that all items are divisible. The adjusted winner procedure was famously applied to resolve the issues of Israel and Palestinians, Panama canal treaty and Camp David accords. There is another program called Family winner, which assists in dividing the assets between the family members or when couples decide to get separated and assets needs to be divided between them.

3.2. Automated technology/tools used for online dispute resolution

The following are the mostly used automated tools for online dispute resolution [10, 11]:

Blind bidding:

These systems invite parties in dispute and ask them to submit their acceptable settlement offers with confidentiality and determine acceptable terms for both the parties.

For example, Smartsettle.

Drafting collaboration:

These are based on tools, which facilitate parties to review draft documents and forms to resolve a dispute amicably.

For example, SettlementIQ, MicroPact.

Automated negotiation:

These systems help in calculating all possible outcomes and also help the parties to arrive at a win-win situation as an outcome of the negotiation process.

For example, Modria, Smartsettle.

Virtual mediation rooms:

It facilitates mediation remotely in real time through videoconferencing.

For example, ADR Group's ADRg Express, Virtual Courthouse, Skype, and Zoom.

Arbitration systems:

It facilitates arbitrators to conduct arbitration process online from different locations, through videoconferencing, and so on.

For example, AAA /DecisionQuest's CaseXplorer Arbitration eQuibbly, Traffic Penalty Tribunal.

Online court case initiation:

Here, disputed parties or their legal representatives file claims and supporting documents through an online tool.

For example, Rolls Building.

Online courts:

Judges do the hearing and pass their judgments on cases by using an official online platform, without the need for face-to-face interaction with disputed parties.

For example, eCourtroom.

Agreement monitoring:

These are based on compliance and monitoring tools, which helps in reporting and analysis of agreements and identifying the breach of agreement.

For example: Rechtwijzer, Our Family Wizard.

4. Online dispute resolution and corporate executives of India

Personal interviews were conducted on 50 senior professionals working in different companies and in different fields like marketing, sales, finance, human resource, software development, and so on in India. These companies belong to the Banking, Consulting, Fast-Moving Consumer Goods (FMCG), Information Technology (IT), and Service sector. From the sample of 50 working professionals, 46 professionals were working in big multinational companies and 4 were entrepreneurs having their own business setup. Each professional interviewed had more than 10 years of working experience and were at the leadership position using online communication most often in their day-to-day professional work. These professionals belonged to the age category of 35–50 years and the sample consists of 31 males and 19 females. During the personal interaction (Interview) with these senior executives of corporate India, they pointed out that most of the conflicts arise because of the misinterpretation/misunderstanding of the written communication through email. Emails are overtly used in the corporate system for communication and is very important to develop the proper email-writing skills [12]. In most of the cases of disputes/conflicts, the other person misunderstands the intended meaning of the email, or it may also happen that the person who is writing the email makes wrong choices of words in the email. The respondents also pointed out that in many instances, it so happen that the email is made c.c. to multiple persons for information sharing and certain message is intended for a particular person but the other person misinterprets the intention and takes it on him and this results in ego clashes and emotional turmoil and affects the relationship of the parties involved. Some executives have suggested that we must take email writing very seriously and use lot of caution before sending it to the recipients.

They further agreed that there are many email exchanges on a daily basis and sometimes people write emails in a hurry without understanding the gravity of the message and without realizing that this may be interpreted in many ways by the recipient and end up committing mistakes. To resolve such kind of disputes they either clarify their point by writing another email and clarify the confusion and if it does not get resolved, then either by making a phone call to the person or by meeting in person to have a face-to-face discussion. For certain issues where things become difficult to manage, they mostly take the help of their senior person who is well respected by both parties and request him to mediate and resolve the issue. To resolve the conflict, mostly they follow "dual concern model" of Black and Mouton [13], where they try to reach to the amicable solution by creating a balance between the needs and expectation of all involved persons, while having concern for self as well as for others. Typically, they follow "cooperation conflict resolution style" more often, which is suggested in the dual concern model. In cooperation conflict resolution style, individuals are assertive for their needs and empathetic also for the other person's needs/expectations/requirements. When the facilitator (senior/colleague) is involved in resolving the dispute, mostly he mediates between involved persons, but sometimes he uses mediation-arbitration approach (combination of mediation and arbitration) as this approach gives more flexibility to resolve the issue. In mediation-arbitration approach, first, the facilitator uses mediation to resolve the issue, but if it does not work, then the facilitator takes the decision and communicates it to the involved persons [14].

It came into light that most of the companies have legal departments within the organizational system that takes care of the dispute cases arising from customers, suppliers, business partners, and competitors. Legal department referred to the dispute/cases where organization's interests/reputation is involved. In the absence of the legal departments, companies take the professional help from dispute resolution agency or through legal professionals. Sometimes legal department also sought the help of dispute resolution agency in case they realized the need for them.

An executive working at a very senior position in a leading airline company has revealed that they get many customer complaints and disputes related to flight delay and deficiency of services while traveling. These unhappy customers put their complaints in consumer court/forum and take legal course. Attending to such cases (court hearings) takes a lot of time and attention. He agreed that online dispute resolution can help his company by saving lot of their bandwidth (time and effort), and there is faster resolution of the customer's issues.

It has been highlighted that, presently, online dispute resolution agencies are not trending in India like European countries, and these agencies are still evolving.

5. Impact of online dispute resolution

Online dispute resolution has provided a very positive impact to corporate, legal professionals (Lawyers) dealing in dispute resolution, ombudsmen and regulators, and court of law. Corporates have to deal with lot of disputes arising from customers, business partners, vendors, competitors, as well as from other business entities.

Corporates have to deal with disputes arising from various geographical locations and dealing with them involves lot of time, money, and manpower, and many a times, their image is also at stake. ODR enables the corporates to resolve these disputes at a faster speed and helps them with out-of-court settlement. Corporates are thus getting benefitted by avoiding expensive litigation and reputation loss by using ODR.

Legal professionals dealing in dispute resolution are getting benefitted by online dispute resolution techniques as it improves their efficiency in resolving the cases, and this makes their clients happy and satisfied.

Ombudsmen and regulators are loaded with a massive number of inquiries and complaints, and they are under tremendous pressure to provide satisfactory resolution within the timeline. Online dispute resolution has certainly eased out their load by providing quick resolution.

Usage of online dispute resolution has certainly eased the load of courtrooms and judiciary and made them more efficient in terms of resolution of disputes.

Users/disputed parties are getting benefitted with faster resolution with the online mode of dispute resolution; however, this system needs to be further invested and developed and make it more available to the users [15].

6. Advantages of online dispute resolution

The online dispute resolution primarily saves lot of travel time and cost, and also negotiation/discussion can take place at the convenience of the concerned parties. Otherwise, in case of face-to-face discussion/dispute resolution, one party may have the location/environment advantage and may influence the resolution process in his favor to a great extent. Online communication reduces the power difference and discomfort of face-to-face confrontation situation. The online dispute resolution offers similar ground for the disputed parties where both will remain in their comfortable zone. It also offers the communication between the concerned parties in the controlled environment. The online dispute resolution process can take into consideration both synchronous as well as asynchronous communication between the parties where communication can take place real time as well as delayed. In case of asynchronous communication, the party will get time to properly phrase his message or prepare a good presentation of his point of view or situation. Online communication not only enables to control your communication but also helps in monitoring the same. Thus, online communication can also be called as hyperpersonal communication [16].

The advantages of the online dispute resolution can be summarized as follows:

Cost: online dispute resolution is often less expensive as compared to the traditional legal process of litigation.

Efficiency: online dispute resolution can often resolve the dispute in lesser time.

Participation and control: disputants using online dispute resolution mode are equally involved in the process to resolve the dispute and exercise more control on the outcome of the dispute.

Flexibility: parties using online dispute resolution can have more flexibility in terms of timing and availability than the traditional legal process. Online dispute resolution can allow parties in different locations or countries to avoid the costs and inconveniences of travel.

7. Limitations of online dispute resolution

Online dispute resolution has certain limitations as it is based on online communication. It has certain disadvantages over dispute resolution done through face-to-face interaction. First, cues of nonverbal communication are missing in online communication which is established through email or chat rooms. This makes ODR less effective in comparison to face-to-face dispute resolution. Nonverbal cues like appearance, dressing, gestures, face expressions, eye contact/expressions, and voice pitch/intensity are very critical to communication and help in making a perception about the person. Nonverbal communication accounts to 90% of the communication, while verbal communication is only 10% [17]. In case of messages communicated through email and chat, the message understanding is only dependent on written message [18].

Truthfulness/genuineness of the person is also in question in case of online communication as it is easier to present false information and deceive in case of online communication (email and chat) rather than face-to-face interaction. As per the interpersonal deception theory given by Buller and Burgoon [19], deception is easier in online communication. Deception is a message, which is knowingly communicated by the sender to generate false belief in the mind of the receiver. They further pointed out that the sender may adopt one of the three strategies to deceive the receiver, such as falsification (presenting a fiction), concealment (hiding the truth), and equivocation (avoiding the issue).

Online communication being web-based communication is very much exposed to online security related to privacy and confidentiality issues. Here, the confidentiality of the discussion can be compromised and some important information can be leaked to unwanted persons. In addition, online dispute resolution is based on the involved person's being technology friendly. Even in the present scenario, there are many people who are not so comfortable with the contemporary communication technology, and they have their own reservations in using it for the communication purposes.

8. Conclusion

It can be concluded that the scope of online dispute resolution is very wide and today more and more people are getting oriented toward this mode of dispute resolution. Online dispute resolution agencies are approached for resolving personal as well as professional disputes and the approach used by these agencies to resolution is quite contemporary and well appreciated by the users. Online dispute resolution provides more satisfaction to the disputed parties as it offers solution with a lot of convenience. This process is certainly faster, cost-effective, and more approachable when compared to the traditional legal system.

In the context of India, however, ODR is quite beneficial by looking at the number of disputes that arises but most of the people are not aware with this online dispute resolution process, and they are largely dependent on the legal system and grievance redressal agencies/forums. Also not many companies are providing online dispute resolution facility unlike Western countries where online mode of dispute resolution is quite popular and easily accessible. By looking at the current scenario and future scope, online dispute resolution is quite wide, but this system needs a lot of improvisation to make it more dependable and approachable/accessible. Online dispute resolution is a quasi-legal process, but strengthening of ODR system will reduce the burden on traditional legal system substantially and will help in faster and more efficient resolution of disputes. As per the ministry of consumer affairs, food, and public distribution of India, there are over 450,000 consumer cases that are pending in 630 consumer courts of the country. Out of the total pending consumer court cases, 40% of these cases are related to the defective products and banking sector, while the remaining 60% are related to insurance, housing deals, financial deals, electricity cases, medical negligence cases, and so on [20, 21].

It may be appropriate for the companies who are facing lot of consumer complaints/disputes to get a tie-up with a dispute resolution agency and suggest/invite consumers to opt for this mode of dispute resolution. This would not only benefit the company by avoiding legal hassles and settling the dispute off-court, but customers will also get benefitted with faster and cost-efficient resolution.

The public perception of ODR shall encourage the use of modern forms and formal legal obstacles should be reduced—offering the opportunity to those who, bearing in mind the limitations of electronic methods, are not afraid to settle disputes by using them (IDABC, 2004) [3].

Author details

Deepak Verma[1]*, Anshu Banwari[1] and Neerja Pande[2]

*Address all correspondence to: deepakverma711@rediffmail.com

1 Jaypee Institute of Information and Technology, Noida, India

2 IIM Lucknow, Noida Campus, Noida, India

References

[1] Abdel Wahab MS, Katsh E, Rainey D, editors. Online Dispute Resolution: Theory and Practice. The Hague: Eleven International Publishing; 2012. p. 357

[2] Lodder AR, Zeleznikow J. Enhanced Dispute Resolution through the Use of Information Technology. Cambridge: Cambridge University Press; 2010

[3] Mania K. Online dispute resolution: The future of justice. International Comparative Jurisprudence. CA, Elsevier B.V.; 2015. pp. 76-86

[4] Suquet M et al. Online Dispute Resolution in 2010: A Cyberspace Odyssey? In: Poblet M, Abrahams B, Zeleznikow J, editors. Proceedings of the 6th international workshop on online dispute resolution. In conjunction with the 23rd international conference on legal knowledge and information systems, Jurix. 2010. pp. 1-12. Retrieved from: http://ceur-ws.org/Vol-684/ODR2010proceedings.pdf

[5] Phillips MJ, Gully SM. Organizational Behaviour: Tools for Success. South-Western: International Edition; 2012

[6] Banwari A, Shakeel M, Verma D. "Does Preference For Negotiation Style Depends On Religious Orientation?", International Journal of Applied Business & Economic Research (IJABER) Part 2. Serial Publishing Pvt. Ltd.; Vol. 15, Issue 23, Pages 79-95. ISSN: 0972-7302

[7] Hindrikis KV, Jonker CM. Special issue on 'human factors and computational models in negotiation. Group Decision and Negotiation. 2012;**21**(1):1-2

[8] Hammond A-MG. How do you write "yes"? A study on the effectiveness of online dispute resolution. Conflict Resolution Quarterly. March 2003;**20**(3):261-286. DOI: 10.1002/crq.25

[9] Taylor FE. Online Dispute Resolution, Conflict & Negotiation. Sage Publication; 2016. pp. 173-192

[10] Online Dispute Resolution and Ombudsmanship [Internet]. 2016. Available from: https://www.mediate.com/pdf/fowlie.pdf [Accessed: Dec 8, 2017]

[11] Brams J, Taylor AD. Fair Division: From Cake Cutting to Dispute Resolution. Cambridge: Cambridge University Press; 1996

[12] The Impact of ODR Technology on Dispute Resolution in the UK [Internet]. 2016. Available from: https://blogs.thomsonreuters.com/legal-uk/wpcontent/uploads/sites/14/2016/10/BLC_ODRwhitepaper.pdf [Accessed: Nov 24, 2017]

[13] Conflict Resolution [Internet]. Available from: https://en.wikipedia.org/wiki/Conflict_resolution via wikipedia [Accessed: Feb 22, 2018]

[14] What are the Different Types of Conflict Resolution? [Internet]. Available from: http://www.wisegeek.com/what-are-the-different-types-of-conflict-resolution.htm [Accessed: Feb 22, 2018]

[15] Verma D, Pande N. An exploratory study of the e-mail writing skills of B-school campus recruits in contemporary corporate India. LBS Journal of Management & Research. 2014;**12**(1):1-5

[16] Walther JB. Computer-mediated communication: Impersonal, interpersonal, and hyperpersonal interaction. Communication Research. 1996;**23**:3-43

[17] Verma D, Pande N. An assessment of nonverbal communication skills of business school campus recruits in India: Implications for stakeholder communication. International Journal of Intelligent Enterprise. 2016;**3**(3/4):251-264

[18] Burgoon JK, Walther JB. Nonverbal expectancies and the evaluative consequences of violations. Human Communication Research. 1990;**17**:232-265

[19] Buller D, Burgoon J. Interpersonal deception theory. Communication Theory. 1996;**6**(3):203-242. DOI: 10.1111/j.1468-2885.1996.tb00127.x

[20] Are Consumer Courts Really Serving the Consumers? [Internet]. Available from: https://economictimes.indiatimes.com/special-report/are-consumer-courts-really-serving-the-consumers/articleshow/7959190.cms [Accessed: Feb 23, 2018]

[21] Over 4.5 Lakh Cases Pending in Consumer Courts of the Country [Internet]. Available from: http://www.newindianexpress.com/nation/2017/aug/08/over-45-lakh-cases-pending-in-consumer-courts-of-the-country-1639896.html [Accessed: Feb 23, 2018]

Chapter 10

Understanding Stress in Communication Management: How It Limits the Effectiveness at Personal and Organizational Level

Fawad Kaiser

Additional information is available at the end of the chapter

http://dx.doi.org/10.5772/intechopen.76527

Abstract

The fundamental challenge for refining theories of stress in communication management is to understand the assumption that stress produces anxiety. Important areas of study in this area include the experience and meaning of stress, the role of appraisal and emotion in anxiety management, and the range of behavioural and psychological responses to stress and anxiety. Applying stress management theories to communication management work practices shows the rise of a culture of stress leading to anxiety and depression. Argument against this point is constant health surveillance, combined with improved methods for screening and monitoring, which virtually guarantees finding something wrong with every individual. Good employment practice includes assessing the risk of stress among employees and involves looking for pressures at work which could cause high and long lasting levels of stress. Deciding who might be harmed by these factors and whether organizations are doing enough to prevent that harm is very important. Therefore it is almost mandatory for every organization to understand and manage these behavioural issues to remain organized. It is not just for the wellbeing of the employees but the organization as a whole.

Keywords: stress, anxiety, communication management

1. Introduction

Stress is defined as interaction between the situation and the individual [1]. In today's communication environment the energy demands from social networking services (SNSs) can result in fatigue and can cause intense emotional and physical strain [2]. More importantly potential stressors at the work place explain the necessity of stress management which impacts the effectiveness of the project teams in a multi-project environment presenting this type of environment as a stress factor [3]. Coping strategies influence how employees' deal with the stress. Maladaptive coping strategies are less influential than adaptive coping strategies and stress management programs directed on enhancing coping strategies have better results [4]. Workplace error management has also been a focus of various studies. Applying the transactional theory of stress frame work King and Beehr has examined the connection of error management strategies to employees' well-being and showed that positive error management (PEM) did not appear to reduce strains, but negative error management (NEM) increased them and difference in the approach between the individual and organization can affect employees' interpretations and management of errors [5]. ICT-based communication is known to be a source of stress. These stressors are accumulative in nature and grow over the space of time to increase stress levels at the end of day [6]. Considering that relationships have a positive influence on managing stress research work shows that couple-targeted and stress-targeted interventions improve relationship among couples which in turn help to improve the health of their children [7].

This study examines how Healthcare communication professionals (HCPs) have to deal with workplace stress while working in a stressful environment. Most of the HCPs are seen to use avoidance – centered coping strategies and Hospital Management need to constantly review their work place stress programs to increase the chances of organizational success [8]. Increased stress and anxiety in the clinical settings are known to have an effect on clinical performance. Programs used with simulations to enhance the coping skills of health profession trainees in reducing stress and anxiety have shown increased levels of effectiveness during high anxiety clinical situations [9]. Effectiveness of a theory-based online intervention to help undergraduate students cope with the stress management has shown shows positive results [10].

2. What is stress?

Stress is a natural phenomenon that can impact an individual both from external and internal world; it can impact his physical or emotional health. And it can affect both of them. Besides, an individual's response to it affects the individual and his environment, both.

Therefore, stress is associated to both internal and external factors. External aspects include physical features like home, work, relationships, routine issues, challenges and expectations etc. Internal aspect is concerned with individual's capability to deal and react with the environmental or (external) elements that induce it. These factors affect the handling capacity of individuals. It includes their health (overall), nutrition, fitness, and emotional well-being, and rest and sleep qualities. Now, stress as a term may be used in many manners and for a various types of reasons. Psychologists argue that there are two varieties of stress: distress and eustress. Distress is harmful while eustress is not harmful. Further, stress has different effects like psychological, behavioural, bio-chemical and physiological stresses, etc.

2.1. Theories on stress

There are many theories on physical processes and external types of stressors. These may be categorized in two ways: "systemic" stress these are based on psychobiology, and physiology [11] and mental stress developed within the field of cognitive psychology [1–14].

2.2. Systemic type (Selye's theory)

In science and media studies, the concept of stress is driven from the works of famous endocrinologist, Hans Selye. According to him, non-specific changes give systemic stress its specific form of reaction. Moreover, writer describes that this stress is 'a state manifested by a mental condition which consists of all the non-specifically induced changes in a biologic system.' It is known as 'General Adaptation Syndrome' (GAS). It is the stereotypical reaction pattern. It occurs in three steps.

(a) There is an alarm reaction. It consists of an initial shock phase, then there is the counter-shock phase. The shock phase shows emotionality. It is autonomous in nature. This releases adrenaline hormone along with gastrointestinal ulcerations. The counter-shock phase initiates the defensive procedures and is branded by amplified adrenocortical activity. (b) If harmful stimulus continues the person reaches the resistance stage. At this step, the alarm reaction signs vanish. This apparently shows that the individual adapts to the stressor. When resistance to this harmful stimulation rises, it drops the other stressor types simultaneously.

(c) If the aversive stimulation continues, it turns in exhaustion. The adaption capacity to the stressor exhausts, the signs of this phase recur, (a) but this time resistance becomes impossible. Irreversible tissue damage occurs. The individual can die if this persists.

2.3. Psychological stress: the Lazarus theory

Two theories are important to any psychological theories. Appraisal: estimate on the person's health and his handling skills. It is based on his attempts and actions to achieve particular demands. Stress is taken as an interpersonal notion. It is neither considered as a type of external stimulation nor a typical physiological, subjective or behavioural pattern. Rather it is seen as a 'shift' between people and their environment.

The handling techniques can be noticed as they lay emphasis on various components of stressful incidents [13]. They may change the person–environment relationship behind stress and negative emotions (problem-focused handling). Also, they can indicate internal elements and support decrease the negative state of emotions. They can also alter the appraisal of the situation (emotion-focused handling).

2.4. Resource theories: a bridge between systemic and cognitive viewpoints

Contrasting to methods discussed above, resource theories are not mainly concerned on stress-creating factors; instead they are concerned with resources that secure the health during stressful encounters. Because self-efficacy and optimism and are the only protective features, self-assurance and coherence signify tripartite methodologies. Self-assurance contains three mixed elements: commitment, self-control and determination unlike threat. And coherence is on trusting the world has a meaning, is benevolent and is predictable. In social support, different forms are examined; for instance, informational, instrumental, emotional and appraisal. During stressful situations, the combating power of people to cope with stress decreases. This damages his capacity to handle stresses further.

This results in spiral loss. The procedure needs to focus on the interaction between demands alter with time. Additionally, this technique reveals it is essential to examine both: the effects of resources and their outcomes.

There are four coping modes. They can be defined as: (a) Individuals who have a higher vigilance rate and lower rates on cognitive avoidance. This is called sensitizers. These individuals are basically focused on reducing the level of uncertainty through leading the attention to stress- related information. (b) People who have opposite patterns are called repressors. These individuals reduce the arousal experience evading aversive evidences. (c) Non-defensives have lower rates on both the aspects. These individuals are thought to adapt flexibility on stressful situations demands. Rather than frequently paying attention to coping strategies such as, vigilance or avoidance. They focus acting in most circumstances. (d) People who exhibit higher scores on both dimensions are called highly nervous.

In vigilant and avoidant coping techniques, the individuals strive reducing both the emotional arousal and subjective uncertainty induced by stressful encounters. In most situations, these two aims are mismatched. The highly-anxious individuals are presumed for expressing fluctuations and so have low efficient coping behavior.

2.5. Transactional model

Richard Lazarus and Susan Folkman, in 1981, suggested stress can result from an "imbalance between demands and resources". It can also occur if "pressure exceeds one's perceived ability to cope". Stress management was premised and developed on the notion that it is not a stressor response rather individual's capability and resource to handle the response. This can be changed, hence making stress controllable.

For developing efficient stress management program, it is extremely important to indicate the characteristics central to control stress, and to indicate the intervention techniques for controlling these factors efficiently. Lazarus and Folkman's stress interpretation is concerned with the relationship between individuals and their external environment (known as the Transactional Model). This model challenges stress as argues it as being harmful. It shows stress becomes positive or a challenging when it not taken as a threat. Moreover, with good stress-handling skills or abilities, stress cannot harm the individual. This model also argues that individuals can learn managing and coping with stress. Their perspective can be altered to empower them with confidence and stress-handling techniques. This can have a good effect on their living. Employees mention different types of stresses. The most common among them are:

2.5.1. Company conflicts

Treatment of bosses/employers/companies with employees.

Job insecurity.

http://dx.doi.org/10.5772/intechopen.76527

Bad company policies.

Unfair work distribution with colleagues.

Communication gap.

Unrealistic assignments.

Obscure expectations.

Tight deadlines.

Overwork.

Rude behaviors.

Inadequate salary/benefits.

Uncomfortable office.

Relationship conflicts.

No cooperation.

Long working hours.

Injustice.

Careless mistakes of subordinates.

2.6. Health realization aka innate health model

Similarly, the said model is established on the notion that it is not crucial to have a potential stressor for individuals who are stressed.

It is not necessary to focus on the stressor therefore. Focusing on stressor appraisal in relation to handling abilities (as in transactional model), this type of model is concerned with the thought procedure of an individual. It argues that it is eventually an individual's thinking which is responsible for obtaining the stress response. As per this model, stress is the appraisal outcome and his circumstances through a mental filter of negativity and insecurity. On the other hand, well-being results from seeing the external environment with a "quiet mind".

For supporting individuals understand the nature of thinking processes, help them to recognize instances of anxious thinking and try to alter them with positive thinking. This helps to reduce stress levels. It occurs when either of these tools stop to work properly or when it becomes difficult to switch appropriately.

It is appraisal or perception of situation which is essential to determine whether or not it creates stress. It is the root of transactional model of stress [14] in which the capability of a person to reduce or prevent stress. It is shown by individual's appraisal of (a) threat in any condition (primary appraisal), and (b) handling skills to deal with threat (appraisal of secondary kind). The appraisals are formed by past incidents to deal with stress. So, impact these and future behavior making the appraisal process, it along with its behavior continuous. Stress can be dealt by altering the manner the situation is appraised (cognitive methods) or given response to (cognitive or behavior-related methods).

2.7. Factors at work that cause stress

The pressures, demands and daily routines of workplace contribute majorly to stress. There are many things linked with it causing many risks. These are workload, and social and company pressure are some of these elements. Particular to any job are: long hours, overload, pressure, complicated tasks, shorter breaks, monotony, no or less facilities and so on.

Conflicting roles and obscure work can also be a cause. The job development possibilities are crucial barriers against it. Whereas, stress occurs when there is no training, no job security and no promotion. There can be two more sources of stress: work relationships and its culture. Unsupportive, demanding or critical managers can cause stress. While, positive team, work, and social environment reduce the stress levels in staff members. Wrong organizational environment such as, "presenteeism" and unpaid work can also cause stress. While involving people in decision-making processes, sharing updates and providing good facilities help lessen stress. Organisational changes are also a main cause. This includes inadequate consultation, relocation, mergers, restructure "downsizing", redundancies and individual contracts etc.

2.8. Empirical review

Systematic review on work characteristics factors related with psychological issues and related absenteeism [16] (Michie and Williams, [15], unpublished data) are: overwork, long hours, work pressure and their effects on staff members are: lack of participation in decision making processes and inability to control work and poor support, obscure or disorganized managing approach.

2.9. Explanatory model

Three of these elements constitute the influential control-demand work-related strain model [17]. As per the model, work-related strain and health risks occur when demands at work are higher but decision freedom is lower (low personal control overwork plus restricted opportunities for developing skills). And the opposite offers encouragement, motivation for learning and achievement. Of these two, decision latitude is considered more important than demand [18]. It was introduced in 1979, and at that time the model was extended for social support as a predictor of job strain [19]. Karasek's model received satisfactory empirical support to offer a beneficial framework for workplace interventions.

2.10. Individual differences

There are individual differences in experiencing and vulnerability to stress. They are more likely to have stress if they do not have material resources like financial security and psychological resources like coping techniques. Stress harms them more if they respond emotionally, are highly competitive but pressed (type-A behavior). The affiliation between well-being and pressures are thought to be inverted U; when pressure is either low or high, the functioning and well-being becomes low (for instance, during unemployment). Different individuals show different forms of inverted U. This shows their different thresholds for stress response. An effective approach to prevent stress at work ensures work fits the individual, instead of attempting to make individuals fit their jobs that do not suit them.

2.11. Interactions between work and home stress

Progressively, individualistic demands at work reach out homes and social lives of employees. Uncertain, long, and unsocial hours, working afar, taking work at home, more responsibility, job relocation and insecurity can all adversely impact family responsibilities and recreational activities. This can affect the quality of life besides work. It is an essential buffer against work-related stress. Additionally, domestic pressures like financial issues, childcare responsibilities, and bereavement and housing problems can also affect an individuals' efficiency at work-place. Therefore, a malicious cycle sets up that causes stress in one's life. It spills over all the places and creates pressures to deal with their life issues.

Women particularly, experience many types of stresses, as they still have more burdens of domestic responsibilities than the opposite sex. Girls are usually low paid, have low statuses, and may often work help fulfill domestic responsibilities. They can suffer harassment and discrimination as well.

2.12. Individual stress management

Most interventions for reducing health risks of stress consist of individualistic and organizational approaches both. Individualistic approaches involve training, one on one psychological services like counseling, clinical and work-related. They must be aimed for helping to improve individuals' skills and resources to help them change their circumstances. These approaches show active coping as well as rest phases (habituation) of model.

There are various training courses that can help develop active coping approaches—for instance, communication skills, assertiveness, problem-solving, time management and effective management. But there are different sources of stress. That a person can perceive as beyond his power. For instance, structure, culture or management style of an organization. It is essential to understand that these management approaches (concentrate on changing the person) without trying to change the sources of stress. Masking these sources can be counter-productive. For instance, trying to think positively or breathe deeply on a stress-inducing situation can temporarily improve one's mood. But it not be permanent and will cause persistent stress in the individual and in others. The major goal of an individuals' approach must be developing those skills and confidence to reduce stress, and not adapting or accepting a stressful situation.

2.13. Organisational stress management

Stress management and prevention needs interventions on organizational level. It is the organization which creates it. This approach is limited to help those who are already experiencing the problem. It is equivalent to apply sticking plaster on injuries than dealing with the root cause of the problem. Another analogy is to try running up the escalator that goes down. These interventions can have a variety that ranges from structural like staffing, scheduling; to physical, and psychological environment for instance, social support, control on work and participation. There is emphasis on organization than individual. As this issue is well-illustrated by the rules used in Scandinavia. The place has an outstanding record of creating safe and healthy working environments [20].

3. Principles of preventing work stress

Conditions at work are adapted to individuals who differ in physical or psychological abilities. Employees are provided with opportunities for participating in designing their work situation and in improving the effects on work. Technology, job content and work organization, formed so the staff members cannot be exposed to mental or physical strains. This can lead to accidents or illnesses. Remuneration forms and work distribution hours are counted here. Work restrictions or excess control on individuals is avoided. Diversity in work, social interaction and cooperation in work is given importance. Conditions of work should offer opportunities for occupational and personal grooming and training and self-determination.

3.1. Assessing risks at workplace must be considered

The extent of harm from stress at the workplace, the extent to which individuals' come across this hazard, number of staff exposed to it.

The analysis for threats at workplace should cover all the sides in management and design as well as its organization and communal context. Avoidance is the priority though but essential measures should be introduced for controlling and reducing the risks and effects of the threat. Details about all these have six stages [21].

3.1.1. Risk identification

Consistently sort the stressors present at the workplace and the working conditions. Examine the exposure extent for employee categories.

3.1.2. Harm assessment

Collect the evidences that stress exposure is related with ill health in different employee groups or wider organization. This must involve a wide range of health issues, like general malaise symptoms or disorders; and different health-related behaviors like drinking, smoking and sickness etc.

3.1.3. Identification of likely risk factors

Find association between stress exposure and the measurement of harm for identifying the risk factors in groups, and to estimate their significance or size etc.

3.1.4. Description of underlying mechanisms

Recognize and explain the possible tools which show stress exposure is linked with harm to health of a particular group or to an organization.

3.1.5. Audit existing management control and employee support systems

Classify and evaluate all the existing management systems to control the stressors and examine experiences and to support the employees who experience issues.

3.1.6. Residual risk recommendations

Take note of the existing managing control and support systems to make recommendations on residual risks linked with risk factors of work stress. Indeed, legislation is required at organizations for addressing and assessing all the risks include mental health also (for instance, the European Commission's framework directive introduced measures for encouraging improvements in health and safety of employees etc.) To create a secure system involves everything from targeting material, equipment to environment and staff members (for example, to ensure good working skills). Also, this needs to be monitored and review all the system for assessing the effectiveness for prevention and control strategies [19].

Though relations between work and psychological health and sickness have been noticed, but evidence-based interventions for reducing these problems are still unlikely or low. Many organizational strategies and training sessions are used in successful interventions for increasing participation in different features such as problem-solving, decision-making, support, feedback and communication etc. [22].

Various studies reveal that skills for mobilizing support at workplace and participation in resolving issues, and make decisions reported better feedback, feeling better at coping and functioning etc. Among those at leaving risks, and undergoing training showed reduction in depression levels [17]. Staff members, who faced organizational changes but who were given skills to manage stress reported decline in stress levels in the body [23].

Employees who learned empathy skills and better communication with each other also demonstrated less resignations and leave instance [24]. Similarly, members who were physically inactive and were given stress management training showed improvement in their perceived coping abilities. Also, those who were given aerobic exercises had improvements in overall well-being. There was decline in complaints about muscle pain, but they also showed reduction in job satisfaction [22].

Staff that took one of the seven training programs highlighting different stress management aspects like—physiological procedures, coping with individuals or interpersonal awareness procedures—had shown reduced anxiety, depression, psychological exhaustion, strain levels immediately after the training. A further reduction level was seen in exhaustion level in 9–16 months' follow up [25].

Also, the employees that were on long-term absence, and were referred earlier to occupational health (in 2–3 months absence period) had a cut down in their sickness absence in 25–40 weeks before work resumption and from 53 to 72 weeks before departure for medical reasons; this led to larger financial savings [26].

Success in prevention and management of stress depends on organizational culture. It should be taken as supportive to action, rather than as individual's weakness. An openness and cooperative culture should be encouraged at workplace than of criticism and blame, because this is essential. To build this culture needs active leadership skills and role models from the top management, development and implementation of stress policy in the organization and effective systems for identifying issues earlier and for reviewing and improving strategies to develop for addressing the issues. Policies and their implementation must be negotiated with relevant groups or unions and other committees (trade union example is a model agreement to prevent stresses at workplace see the Manufacturing, Science and Finance Union guide) [27].

Lastly, different interventions require good evaluations, so their effectiveness can be examined. Preferably, the technique to achieve this goal must involve higher response rate, reliable measures and a control group. Two techniques which offer comprehensive stress analysis and these are widely applied are the Job Content Questionnaire. This involves measures of job strain predictors as discussed earlier [28] and the Occupational Stress Indicator [29].

3.2. Stress prevention and resilience

Even though many methods are traditionally developed for dealing with the stress outcomes, a lot of research is conducted on stress prevention. It is a subject that is closely related with psychological resilience building. There are a number of "do it yourself" approaches too on this and on resilience-building. They are mainly based on cognitive-behavioural therapy theory and practice.

4. Stress management programs

Nowadays, many companies have started using stress management training programs for their staff members suffering from it. They use different ways for this. Some organizations offer special tools for adapting to this for instance, balls, color diaries etc. Some offer individualistic interventions. This involves stressor monitoring in the patient. Understanding the causes, addressing the stressors and finding out solutions to it. Social support is crucial to it. Society plays a good role to help cope with stress. It has proven highly practical. And trying to avoid stressors is also a way to this. However, it is difficult to avoid it at the workplace especially. Also, altering one's behavioural pattern can help keep away stress.

There are many staff assistance programs. They include different in-house counseling programs. Evaluative research is conducted on EAPs. It educates the individual to control stress and inoculation approaches for instance, biofeedback, cognitive restructuring, and relaxation. Studies prove that such programs may help remove physiological awakening related to higher stress levels. Individuals who master cognitive and behavioural stress-relieving approaches show reduction in tension, fewer sleep issues and better coping with work-related stressors.

Another condition to reduce stress at workplace is simply by altering the workload of a member. Some of them can be too overwhelmed about it to get it done, or some can have so less work that they may not be sure about work about it. Recuperating communication between staff members seems to be a simple technique, but it is highly effective to help decrease stress levels. Moreover, trying to make staff members feel important part of the organization. For instance, offering them a voice in important circumstances prove that the company trusts and value them. To have all the staff members mingle together is a highly motivating and helps shun workplace stress. When the staff members have understanding and share their feedbacks with each other, stress minimizes. Finally, altering the physical conditions at work can also help minimize stress levels. For instance, changing facilities at organization and in environment such as temperature, air, odor and equipment. Intervention can be broken in three steps: primary, secondary and tertiary.

The first step emphasis on removing the stressors altogether. The second step is about assessing stress and finding solutions to handle it by working on the management skills. The third step deals with rehab and recovery of the stress. These stages are considered the most effective in dealing with stress, at work or even outside.

5. Stress and communication process

A current time organization is considered exposed to stressful setting, naturally. The new financial down turn has added fuel to fire to this condition. This is shown by the following case which involves a stressful condition in an 'Accounting & Auditing' company. Call it company A. The said company is known for its huge size and have accomplished greater heights in a shorter time period. Technology is the main source of communication in company A across all levels. It strongly believes to save time and effort using technology. The adjectives suiting the overall atmosphere and employees at company A include demanding, dynamic, have higher expectations to target achieving, shorter deadlines etc. Company A, as most other big guns in technology sector involves unpredictable working hours. And so, it is also famous for comparatively higher monetary compensations it offers for its employees.

Recently, company A undergoes alterations their work policies to fit the changing requirements of their environment. Parallel to this, it is observed that there is an increase in the employee numbers who are leaving this organization and there is more absence and leaves while there is decline in employee performance. This increase in staff turnover is worrisome for the management. They understood that it is crucial to give attention to such issues which dangers their reputation and status in the marketplace.

Under these circumstances, one may understand that the staff members suffer from many types of behavioural problems. However, consider this also that there is increase in offering importance to managing their problems efficiently. As hinted above, the root cause of the case study analysis it for stress in the employees. And also it is related to another issue that is the communication procedure. This shall be done by describing and applying the better-suited theory in relation with the organizational behavior. Following this are the recommendations and conclusion to help managing their issue effectively.

6. Relationship between stress and communication process

In the start, one should understand and try to identify whether the company and its staff members, in particular, suffer from stressful situation.

Stress is defined as "a response, consequence of any action, situation or event that places special demands on a person; and all such actions/situations/events that places special demands are called Stressors" (Gibson, Ivancevich, Donnelly and Konopaske).

7. Effective communication process

Companies requires efficient face-to-face communication system. This type of communications helps reducing stress levels and work dissatisfaction. Whole organization should have good communicating processes. It is one of the most essential company procedures which helps erase many confusions, conflicts and misunderstandings between individuals and the company.

There are three main issues. These include: omission, distortion and message overload. When the sender is frightened by the receiver, message along with interpretation can be affected. The major reflections that crop up with this can be: sexual attraction, social status, and differences etc. These issues lead to message distortions. The communication medium plays an important role in this process. Any inappropriate medium/channel leads to ambiguity, confusion and distortions. To filter the right information and listening selectively are the two main procedures here, when receiving and sending messages, respectively.

8. Measures

Digital technology is an important part of this process. It must be monitored in the organization, regularly. It is because unexpected break can affect the whole procedure, which can lead to stress.

Regular meetings with team also provide solutions to many problems that leads to higher job satisfaction and hence, reduces stress. Similarly, offering opportunities to better communication between team members from the other competency group can lead to higher cooperation in the company.

Create environment that offers two-way communications to employees and the management. All of them should be involved and the management should be approachable and nice to them. Non-verbal communication is equally essential here for creating a good impact on the message receiver.

Interpersonal communication which is the face to face, group meetings and interaction are also essential than emails. So, these strategies must be applied when required.

Feedback and upward communication in this process are most of the times, neglected. Emphasis must be offered to them to reassure the message correctness (message that is given must be understood in its right sense i.e. as it is intended by the sender).

9. Conclusion

In an individual and for an organization, communication, performance, quality of life and stress are extremely interdependent. It is proved that positive communication in company and job satisfaction level are the strong forecasters of stress levels and cooperation breaches. The research on organizational behavior is highly important not only for addressing performance issues; but also saving job costs in the organization. Being effective and efficient are the guidelines of all the organizations. Organization behavioural studies assist the authorities to realize that individuals perform better, their productivity increases, customers become more satisfied and competitive position improves with this. So, it is important and mandatory for all organizations for understanding and managing such issues, and avoid being disorganized. It is not only important for the betterment of the staff members but also for the organization in the long run.

Author details

Fawad Kaiser

Address all correspondence to: fawad_shifa@yahoo.com

Shifa Tameer E-Millat-University, Shifa International Hospital, Islamabad, Pakistan

References

[1] Michie S. Causes and management of stress at work. Occupational and Environmental Medicine. 2002;**59**:67-72

[2] Lee AR, Son S-M, Kim KK. Information and communication technology overload and social networking service fatigue: A stress perspective. Computers in Human Behavior. 2016;**55**:51-61

[3] Bresnen M, Edelman L, Newell S, Scarbrough H, Swan J. Social practices and the management of knowledge in project environments. International Journal of Project Management. 2003;**21**:157-166

[4] Holton MK, Barry AE, Chaney JD. Employee stress management: An examination of adaptive and maladaptive coping strategies on employee health. 2015;**53**(2):299-305. DOI: 10.3233/WOR-152145

[5] King BS, Beehr TA. Working with the stress of errors: Error management strategies as coping. International Journal of Stress Management. 2017;**24**(1):18-33

[6] Reinke K, Gerlach G, Tarafdar M, Stock RM. ICT-based communication events as triggers of stress: a mixed methods study. In: International Conference on Information Systems(ICIS). Dublin Proceedings. AIS Electronic Library; 2016

[7] Lavner JA, Bradbury TN. Protecting relationships from stress. Current Opinion in Psychology. 2017;**13**:11-14

[8] Gilstrap CM, Bernier D. Dealing with the demands: Strategies healthcare communication professionals use to cope with workplace stress. Qualitative Research Reports in Communication. 2017;**18**(1):73-81. DOI: 10.1080/17459435.2017.1330277

[9] Hintz S, Frazier PA, Meredith L. Evaluating an online stress management intervention for college students. Journal of Counseling Psychology. Apr 2015;**62**(2):137-147

[10] Ignacio J, Dolmans D, Scherpbier A, et al Stress and anxiety management strategies in health professions' simulation training: A review of the literature BMJ simulation and technology enhanced learning published online first: 06 April 2016

[11] Selye H. Stress in Health and Disease. Boston, MA: Butterworth's, Inc; 1976

[12] Lazarus RS. Theory-based stress measurement. Psychological Inquiry. 1990;**1**:3-13

[13] McGrath JE. Methodological problems in research on stress. In: Krohne HW, Lane L, editors. Achievement, Stress and Anxiety. Washington, DC: Hemisphere; 1982; pp. 19-48

[14] Lazarus S, Folkman S. A comprehensive account of the transactional model of stress and its management. In: Stress Appraisal and Coping. New York: Springer; 1984

[15] Michie S, Williams S. Reducing work related psychological ill health and sickness absence: A systematic literature review; Occopational and Environmental Medicine. 2003;**60**:3-9

[16] Williams S, Michie S, Pattani S. Improving the health of the NHS workforce: Report of the partnership on the health of the NHS workforce. Research report. Nuffield Trust; 1998

[17] Karasek RA, Theorell T. A seminal book that combines a clear theoretical and empirical approach to work stress. In: Healthy Work: Stress, Productivity, and the Reconstruction of Working Life. New York: Basic Books; 1990

[18] Johnson JV, Stewart W, Friedlund P, et al. Long-term psychosocial work environment and cardiovascular mortality among Swedish men. American Journal of Public Health. 1996;**86**:324-331

[19] Hall EM, Johnson JV. Depression in unemployed Swedish women. Social Science & Medicine. 1985;**27**:1349-1355

[20] Gardell B, Gustavsen B. Work environment research and social change: Current developments in Scandinavia. Journal of Occupational Behaviour. 1980;**1**:3-17

[21] Health and Safety Executive. A thorough guide as to how to apply a risk management approach to work stress. In: Organisational Interventions for Work Stress: A Risk Management Approach. Norwich: Her Majesty's Stationery Office; 2000

[22] Gronningsaeter H, et al. Improved health and coping by physical exercise or cognitive behavioral stress management training in a work environment. Psychology and Health. 1992;**7**:147-163. CrossRefWeb of Science

[23] Cox T, Griffiths A. The nature and measurement of work stress: Theory and practice. In: Wilson JR, Corlett EN, et al., editors. Evaluation of Human Work: A Practical Ergonomics Methodology. 2nd ed. London: Taylor & Francis; 1995. pp. 783-803

[24] Smoot SL, Gonzales JL. Cost-effective communication skills training for state hospital employees. Psychiatric Services 1995;**46**:819-822. PubMedWeb of Science

[25] Kagan NI, Kagan H, Watson MG. Stress reduction in the workplace: The effectiveness of psychoeducational programs. Journal of Counseling Psychology. 1995;**42**:71-78. CrossRefWeb of Science

[26] Malcolm RM, Harrison J, Forster H. Effects of changing the pattern of referrals in a local authority. Occupational Medicine. 1993;**43**:211-215. Abstract/FREE Full Text

[27] Manufacturing, Science and Finance Union. Preventing Stress at Work: An MSF Guide. London: MSF; 1997

[28] Karasek R, Brisson C, Kawakami N, et al. The job content questionnaire (JCQ): An instrument for internationally comparative assessment of psychosocial job characteristics. Journal of Occupational Health Psychology. 1988;**3**:322-355

[29] Cooper CL, Sloan SJ, Williams S. Occupational Stress Indicator Management Guide. NFER-Nelson: Windsor; 1998